Roger Lee (Ed.)

Software Engineering Research, Management and Applications

Studies in Computational Intelligence, Volume 150

Editor-in-Chief
Prof. Janusz Kacprzyk
Systems Research Institute
Polish Academy of Sciences
ul. Newelska 6
01-447 Warsaw
Poland
E-mail: kacprzyk@ibspan.waw.pl

Further volumes of this series can be found on our homepage: springer.com

Vol. 128. Fatos Xhafa and Ajith Abraham (Eds.)
Metaheuristics for Scheduling in Industrial and Manufacturing Applications, 2008
ISBN 978-3-540-78984-0

Vol. 129. Natalio Krasnogor, Giuseppe Nicosia, Mario Pavone and David Pelta (Eds.)
Nature Inspired Cooperative Strategies for Optimization (NICSO 2007), 2008
ISBN 978-3-540-78986-4

Vol. 130. Richi Nayak, Nikhil Ichalkaranje and Lakhmi C. Jain (Eds.)
Evolution of the Web in Artificial Intelligence Environments, 2008
ISBN 978-3-540-79139-3

Vol. 131. Roger Lee and Haeng-Kon Kim (Eds.)
Computer and Information Science, 2008
ISBN 978-3-540-79186-7

Vol. 132. Danil Prokhorov (Ed.)
Computational Intelligence in Automotive Applications, 2008
ISBN 978-3-540-79256-7

Vol. 133. Manuel Graña and Richard J. Duro (Eds.)
Computational Intelligence for Remote Sensing, 2008
ISBN 978-3-540-79352-6

Vol. 134. Ngoc Thanh Nguyen and Radoslaw Katarzyniak (Eds.)
New Challenges in Applied Intelligence Technologies, 2008
ISBN 978-3-540-79354-0

Vol. 135. Hsinchun Chen and Christopher C. Yang (Eds.)
Intelligence and Security Informatics, 2008
ISBN 978-3-540-69207-2

Vol. 136. Carlos Cotta, Marc Sevaux and Kenneth Sörensen (Eds.)
Adaptive and Multilevel Metaheuristics, 2008
ISBN 978-3-540-79437-0

Vol. 137. Lakhmi C. Jain, Mika Sato-Ilic, Maria Virvou, George A. Tsihrintzis, Valentina Emilia Balas and Canicious Abeynayake (Eds.)
Computational Intelligence Paradigms, 2008
ISBN 978-3-540-79473-8

Vol. 138. Bruno Apolloni, Witold Pedrycz, Simone Bassis and Dario Malchiodi
The Puzzle of Granular Computing, 2008
ISBN 978-3-540-79863-7

Vol. 139. Jan Drugowitsch
Design and Analysis of Learning Classifier Systems, 2008
ISBN 978-3-540-79865-1

Vol. 140. Nadia Magnenat-Thalmann, Lakhmi C. Jain and N. Ichalkaranje (Eds.)
New Advances in Virtual Humans, 2008
ISBN 978-3-540-79867-5

Vol. 141. Christa Sommerer, Lakhmi C. Jain and Laurent Mignonneau (Eds.)
The Art and Science of Interface and Interaction Design (Vol. 1), 2008
ISBN 978-3-540-79869-9

Vol. 142. George A. Tsihrintzis, Maria Virvou, Robert J. Howlett and Lakhmi C. Jain (Eds.)
New Directions in Intelligent Interactive Multimedia, 2008
ISBN 978-3-540-68126-7

Vol. 143. Uday K. Chakraborty (Ed.)
Advances in Differential Evolution, 2008
ISBN 978-3-540-68827-3

Vol. 144. Andreas Fink and Franz Rothlauf (Eds.)
Advances in Computational Intelligence in Transport, Logistics, and Supply Chain Management, 2008
ISBN 978-3-540-69024-5

Vol. 145. Mikhail Ju. Moshkov, Marcin Piliszczuk and Beata Zielosko
Partial Covers, Reducts and Decision Rules in Rough Sets, 2008
ISBN 978-3-540-69027-6

Vol. 146. Fatos Xhafa and Ajith Abraham (Eds.)
Metaheuristics for Scheduling in Distributed Computing Environments, 2008
ISBN 978-3-540-69260-7

Vol. 147. Oliver Kramer
Self-Adaptive Heuristics for Evolutionary Computation, 2008
ISBN 978-3-540-69280-5

Vol. 148. Philipp Limbourg
Dependability Modelling under Uncertainty, 2008
ISBN 978-3-540-69286-7

Vol. 149. Roger Lee (Ed.)
Software Engineering, Artificial Intelligence, Networking and Parallel/Distributed Computing, 2008
ISBN 978-3-540-70559-8

Vol. 150. Roger Lee (Ed.)
Software Engineering Research, Management and Applications, 2008
ISBN 978-3-540-70774-5

Roger Lee
(Ed.)

Software Engineering Research, Management and Applications

Prof. Roger Lee
Computer Science Department
Central Michigan University
Pearce Hall 413
Mt. Pleasant, MI 48859
USA
Email: lee1ry@cmich.edu

ISBN 978-3-642-08967-1 e-ISBN 978-3-540-70561-1

DOI 10.1007/978-3-540-70561-1

Studies in Computational Intelligence ISSN 1860949X

Typeset & *Cover Design:* Scientific Publishing Services Pvt. Ltd., Chennai, India.

Printed in acid-free paper

9 8 7 6 5 4 3 2 1

springer.com

Preface

The 6th ACIS International Conference on Software Engineering, Research, Management and Applications (SERA 2008) was held in Prague in the Czech Republic on August 20–22. SERA '08 featured excellent theoretical and practical contributions in the areas of formal methods and tools, requirements engineering, software process models, communication systems and networks, software quality and evaluation, software engineering, networks and mobile computing, parallel/distributed computing, software testing, reuse and metrics, database retrieval, computer security, software architectures and modeling. Our conference officers selected the best 17 papers from those papers accepted for presentation at the conference in order to publish them in this volume. The papers were chosen based on review scores submitted by members or the program committee, and underwent further rounds of rigorous review.

In chapter 1, Luciana Akemi Burgareli et al. discuss the challenges of managing variability in software development product lines. The authors present a strategy for variability management, and perform a case test of that strategy on an existing product line of Brazilian Satellite Launcher Vehicle software.

In chapter 2, Ivan Garcia et al. look into the use of questionnaire-based appraisal. They then apply this method to business models, and recommend for use with small and medium sized enterprises.

In chapter 3, Abbas Heydarnoori investigates the challenges of the software deployment process. He proposes that certain automatic tools and techniques be used to streamline the deployment process, and he then provides a general overview of these tools and techniques.

In chapter 4, James H. Hill and Aniruddha Gokhale present a simple technique, based on baseline profiling, for searching or deployments of both distributed real-time and embedded systems that improve performance along a simple quality-of-service concern. They perform experiments to verify the validity of their technique, and present the observations of those experiments to recommend new deployments.

In chapter 5, Lucia Kapová et al. address the problems associated with the common methods of transforming Business Process Modeling Notation (BPMN) into Business Process Executable Language (BPEL), as related to Service-oriented Architecture. The authors propose an enhancement of the existing transformation algorithm which provides a complete transformation, while preserving a larger part of the intent of the original BPMN descriptions that previously possible.

In chapter 6, Gang Huang et al. investigate the challenges in Software Architecture (SA) associated with the flattening of hierarchical SA models in order to transform platform-independent models into platform-specific models. The authors note the problem of lost comprehensibility, redundancy and consistency in the transformation process, and recommend a new systematic approach that preserves these qualities.

In chapter 7, Otso Kassinen et al. present a case study that provides practical guidelines for Simbian OS software development based on a three-year mobile software research project on the criterion of networking middleware and collaborative applications. The authors present observations that advocate the use of platform-independent solutions, the minimization of project dependencies and the representation of complex activities in human-readable form.

In chapter 8, Muhammad Bilal Latif and Aamer Nadeem investigate the extraction of a Finite State Machine (FSM) as associated with the writing of requirements specifications for safety-critical systems. They discuss the challenge of automatic FSM generation from Z-specifications as caused by difficulties in identifying and extracting pre- and postconditions. The authors then present an automated approach to the separation of this data, as a solution to this problem, and provide tools and experimentation to support their suggestion.

In chapter 9, Tomás Martínez-Ruiz et al. suggest a SPEM extension that will support the variability implied in a software process line. They provide for new methods in their extension, in order to allow for the variability needed in a software process line.

In chapter 10, Noor Hasnah Moin and Huda Zuhrah Ab. Halim propose a hybrid Genetic Algorithm to determine the order of requests to be scheduled in the data routing of a telecommunications network. The authors then discuss the design of three algorithms based on the Variable Neighbor Search. They conclude by comparing the performance of these algorithms on a set of data from the OR library.

In chapter 11, Iulian Ober and Younes Lakhrissi propose the use of events as a first class concept for the compositions of software components. They show how the approach can be applied to any language based on concurrent opponents and illustrate their claim with examples.

In chapter 12, Annie Ressouche et al. investigate the challenges of creating automatic specification and verification tools for synchronous languages. They design a new specification model based on a reactive synchronous approach. In practice they design and implement a special purpose language that allows for dealing both with large systems and formal validation.

In chapter 13, Haldor Samset and Rolv Bræk readdress the notion of active services in the context of Service-oriented Architecture. In their paper, the authors explain how active services and their behaviors can be described for publication and discovery.

In chapter 14, Ilie Şavga and Michael Rudolf show how the use of a history of structural component changes enables automatic adaptation of existing adaptation specifications.

In chapter 15, Dimitrios Settas and Ioannis Stamelos propose the Dependency Structure Matrix (DSM) as a method that visualizes and analyzes the dependencies between related attributes of software project management antipatterns in order to reduce complexity and interdependence. The authors exemplify their solution with a

DSM of 25 attributes and 16 related software project management antipatterns that appear in the literature on the Web.

In chapter 16, Gang Shen proposes a practical curriculum for an embedded systems software engineering undergraduate program. The curriculum advocates a proactive learning setting in close cooperation with the relevant industry.

In chapter 17, Yoshiyuki Shinkawa proposes a formal model verification process for UML use case models. The author then describes the results of a test example performed on a supermarket checkout system.

It is our sincere hope that this volume provides stimulation and inspiration, and that it be used as a foundation for works yet to come.

May 2008 Roger Lee

Contents

List of Contributors

Rolv Bræk
Norwegian University of Science, Norway
rolv@item.ntnu.no

Tomáš Bureš
Charles University, Czech Republic
bures@dsrg.mff.cuni.cz

Luciana Akemi Burgareli
Aeronautics and Space Institute, Brazil
luciana@iae.cta.br

Mauricio G.V. Ferreira
Institute for Space Research, Brazil
mauricio@ccs.inpe.br

Daniel Gaffé
University of Nice Sophia Antipolis CNRS, France
daniel.gaffe@unice.fr

Ivan Garcia
Technological University of the Mixtec Region, Mexico
ivan@mixteco.utm.mx

Félix García
University of Castilla-La Mancha, Spain
felix.garcia@uclm.es

Aniruddha Gokhale
Vanderbilt University, USA
a.gokhale@vanderbilt.edu

Huda Zuhrah Ab. Halim
University of Malaya, Malaysia
noor_hasnah@um.edu.my

Erkki Harjula
University of Oulu, Finland
erkki.harjula@ee.oulu.fi

Abbas Heydarnoori
University of Waterloo, Canada
aheydarnoori@uwaterloo.ca

James H. Hill
Vanderbilt University, USA
j.hill@vanderbilt.edu

Petr Hnětynka
Charles University, Czech Republic
hnetynka@dsrg.mff.cuni.cz

Gang Huang
Peking University, China
huanggang@sei.pku.edu.cn

Lucia Kapová
Charles University, Czech Republic
kapova@dsrg.mff.cuni.cz

Otso Kassinen
University of Oulu, Finland
otso.kassinen@ee.oulu.fi

Timo Koskela
University of Oulu, Finland
timo.koskela@ee.oulu.fi

Younes Lakhrissi
Université de Toulouse - IRIT, France
iulian.ober@irit.fr

Muhammad Bilal Latif
Mohammad Ali Jinnah University, Pakistan
m.bilal.latif@gmail.com

Tomás Martínez-Ruiz
University of Castilla-La Mancha, Soain
tomas.martinez@uclm.es

Hong Mei
Peking University, China
meih@sei.pku.edu.cn

Selma S.S. Melnikoff
Polytechnic School of University of Sao Paulo, Brazil
selma.melnikoff@poli.usp.br

Noor Hasnah Moin
University of Malaya, Malaysia
noor_hasnah@um.edu.my

Aamer Nadeem
Mohammad Ali Jinnah University, Pakistan
aamern@acm.org

Iulian Ober
Université de Toulouse - IRIT, France
iulian.ober@irit.fr

Carla Pacheco
Technological University of the Mixtec Region, Mexico
leninca@mixteco.utm.mx

Mario Piattini
University of Castilla-La Mancha, Spain
mario.piattni@uclm.es

Annie Ressouche
INRIA Sophia Antipolis - Méditerranée, France
Annie.Ressouche@sophia.inria.fr

Valérie Roy
CMA Ecole des Mines Sophia Antipolis, France
vr@cma.ensmp.fr

Michael Rudolf
Technische Universität Dresden, Germany
s0600108@inf.tu-dresden.de

Haldor Samset
Norwegian University of Science, Norway
haldors@item.ntnu.no

Ilie Şavga
Technische Universität Dresden, Germany
is13@inf.tu-dresden.de

Dimitrios Settas
Aristotle University of Thessaloniki, Greece
dsettas@csd.auth.gr

Gang Shen
Huazhong University of Science and Technology, China
gang_shen@yahoo.com

Yoshiyuki Shinkawa
Ryukoku University, Japan
shinkawa@rins.ryukoku.ac.jp

Ioannis Stamelos
Aristotle University of Thessaloniki, Greece
dsettas@csd.auth.gr

Pavel Sumano
Technological University of the Mixtec Region, Mexico
sumanoop@mixtli.utm.mx

Yanchun Sun
Peking University, China
sunyc@sei.pku.edu.cn

Jie Yang
Peking University, China
yangj@sei.pku.edu.cn

Mika Ylianttila
University of Oulu, Finland
mika.ylianttila@ee.oulu.fi

A Variability Management Strategy for Software Product Lines of Brazilian Satellite Launcher Vehicles

Luciana Akemi Burgareli, Selma S.S. Melnikoff, and Mauricio G.V. Ferreira

Aeronautics and Space Institute - IAE
São José dos Campos, SP, Brazil
luciana@iae.cta.br

Summary. The Product Line approach offers to the software development benefits such as savings, large-scale productivity and increased product quality. The management of variability is a key and challenging issue in the development of the software product line and product derivation. This work presents a strategy for the variability management for software product line of Brazilian Satellite Launcher Vehicles. After modeling the variability, extracting them from use case diagrams and features, the proposed strategy uses a variation mechanism based on a set of Adaptive Design Patterns as support in the creation of variants. The proposed strategy uses as case study the software system of an existing specific vehicle, the Brazilian Satellite Launcher (BSL).

1 Introduction

The goal of product-line engineering is to support the systematic development of a set of similar software systems by understanding and controlling their commonality, which is a quality or functionality that all the applications in a product family share, and variability, which represents a capability to change or customize a system [11].

The software product line engineering may be split into two main activities [15]:

(1) Domain Engineering, also known as Core Assets Development, is the product line activity that encompasses domain analysis, design and implementation. During those phases the core asset that consists of commonality and variability, is constructed to be used in the production of more than one product in product line.
(2) Application Engineering, also called Product Development, is the activity in which the application of the product line is built through the reuse of artifacts created during the Domain Engineering phase [13, 15].

The need for the development of new products leads to the evolution of the product line and that results in increased variability, and, consequently, managing the variability becomes more difficult. A way to make variability management easier and make the variabilities feasible is the modeling of variabilities.

This work presents a strategy whose purpose is to establish a variability management process for product line. Features are extracted from the system through a use case model, and then a model of such features is constructed to help in the categorization and modeling of the variability. After modeling the variabilities, the proposed

R. Lee (Ed.): Soft. Eng. Research, Management & Applications, SCI 150, pp. 1–14, 2008.
springerlink.com

strategy uses a variation mechanism based on a set of Adaptive Design Patterns as support in the creation of variants.

A software of an existing specific launcher vehicle, the Brazilian Satellite Laucher (BSL), is used as case study for the proposed strategy. The BSL is a conventional launcher vehicle whose function is to launch to low altitude orbits, between 200 and 1000km, data collecting and remote sensing satellites, weighing between 100 and 200kg [1].

The proposed strategy intends to facilitate analysis and choice of software reusable solutions to enable the derivation of other vehicles, considered members of the Brazilian Satellite Launcher Vehicles family.

In section 2 of this work the need for a product line approach for the development of the BSL software is justified. In section 3 the proposed variability management strategy is presented. Section 4 shows an example of usage of the proposed strategy, applying it to BSL software. Section 5 presents final considerations.

2 Motivation

In nearly 15 years of BSL software development, the project team has seen a high rate of qualified personnel rotation. Unfortunately, few developers have remained in the project since its inception. Therefore, it is necessary to ensure the understanding of clear documentation and updated models for the launcher, so that there are no difficulties in passing project information to new team members and all people involved can understand the intentions of original developers.

In addition to that, problems are also caused by the obsolescence of used technologies, such as methodology and CASE tools. The BSL software was originally designed using the technologies available in the early 90's. Today, it is ever more difficult to keep the infrastructure for such environments, making it difficult to capture, edit and update models and documents they generated. Therefore, a new development approach, which uses more recent technologies, improves understanding and helps the team to use and update models, is required.

Another important point is that during the long development process, hardware, operational system, compilers and sensors platforms underwent upgrades to new technologies, a trend in any project. Such changes may lead to partial losses (models, architecture and code) or total loss of the software system. Today, to keep up with BSL development it is necessary to develop or adapt software models specific for each launcher version, and that is a high-cost task. That evidences the need for the improvement of software development which allows increased reuse of BSL software models, preventing redundant efforts in the development of new systems, even in face of changes in technologies.

The perspective of new projects that provide for a gradual development of launchers, such as the one announced in 2005 by the Brazilian General Command for Aerospace Technology (CTA) and the Brazilian Space Agency (AEB) [1, 12], also creates the need for a more planned development of software, with the controlled generation of models which can be reused by the other satellite launcher vehicles.

Updating of models and documentation, need for quality reuse and perspectives for new projects are the motivation behind the decision for the application of a product line approach for the BSL software development. The goal is to favor the reuse and enable an easier and faster development of software for Brazilian satellite launcher vehicles, ensuring better understanding of the domain.

We will limit this first work to presenting the variability management strategy for the product line for the Brazilian launcher vehicles software. Therefore, only the Domain Engineering activities required for the creation of artifacts to be used as input to the proposed variability management will be developed. Application of product line for the Brazilian launcher vehicles software, using the variability management strategy, will be presented in future works.

3 Proposed Variability Management Strategy

Figure 1 depicts the first Domain Engineering analysis activities and the activities that compose the proposed variability management strategy.

The presented Domain Engineering activities are responsible for the development of some artifacts which compose the core asset and which are required for the variability management. The first analysis activity of the Domain Engineering is the making of a textual description of the problem in order to obtain the system requirements. System functional requirements are then described in use case diagrams.

The representation of use cases is based on the one suggested by Gomaa [4] which in his product line development process, based on UML, called Evolutionary Software Product Line Engineering Process (ESPLEP), uses stereotypes to distinguish between

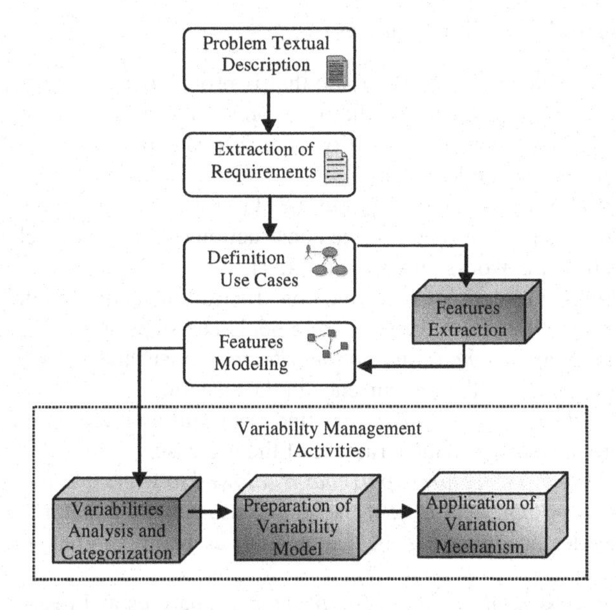

Fig. 1. Activities of Proposed Variability Management Strategy

the different types of use cases. Thus, the proposed strategy uses the stereotypes <<common>> for the use cases which are always used; <<optional>> for the cases used sometimes and <<alternative>> for the cases in which a choice is to be made.

Figure 2 shows an example of a use case diagram for the BSL software product line. After identification of use cases, the next activity to be performed is the extraction of functional features from the constructed use case model.

The next sections detail extraction of features from the use cases and other phases of strategy presented in Figure 1.

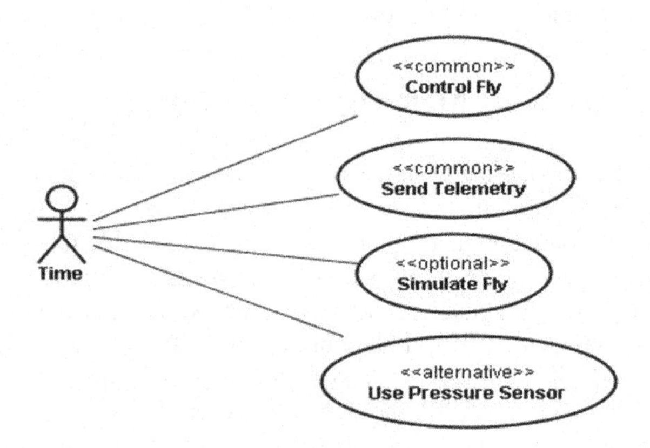

Fig. 2. Example of Use Cases for the BSL

3.1 Extracting and Modeling Features

A feature is a requirement or characteristic that is provided by one or more members of the product line [4]. According to the work presented in [5], a feature model captures the commonality and the variability of a software product lines by means of features and their interdependencies.

Literature shows some diversification in the classification of features and representation of their models [3, 4, 7, 17]. For the construction of feature model, the proposed strategy uses as base the works presented in [4, 7].

Griss [7] defines, in his process for the construction of feature model, the steps to extract functional features from the use case model. The steps are:

Identify mandatory (common) and optional features: lists and classify use cases according to the frequency of their occurrence in the system.

Decompose features into sub-features: finds out sub-features that compose each identified feature, according to the structure of the use case.

Identify variation Point-features (part that is allowed to vary) and their variants (realization of variationPoint-features): if the source use case of a feature contains variation points, consider decomposing the feature into as many sub-features as variation points.

Develop robustness analysis in the feature model: analyses and restructures the feature model as a whole, introducing certain restrictions (mutual_exclusion, requires).

It is possible to incorporate other observations which facilitate extraction of features from use case model, with the following analysis:

- List use cases which may be considered as fundamental to the system. For example, in the case of the launcher vehicle control system, they are the functions which every vehicle needs to perform a control. Those features are candidates for addition to the common features list.
- List secondary use cases, which are not considered fundamental to the accomplishment of the main system purpose. Probably, these functions are represented by a sequence of interactions isolated from the main system function. (i.e. test simulation module). Those features are candidates for addition to the optional features list.
- List among use cases which generate fundamental and secondary features listed, variants, represented by groups of requirements which may be necessary alternately between them. For example, to send data to the ground station, the system needs the measurements read by sensors or calculated. Thus, variants may be alternative, i.e. measurement of pressure sensor or altitude calculation measurement. Such features are candidates for addition to the list of variationPoint_features and variants.

Analysis based on use cases facilitates the construction of the feature model. Figure 3 illustrates an example of a feature model for the BSL, constructed according to the analysis previously described and using stereotypes UML to distinguish the feature types, as suggested in [4].

The feature model is important for the product line approach because it makes distinctions between commonality and variability easier and clearer, helping developers to identify which parts of the system need to support variability.

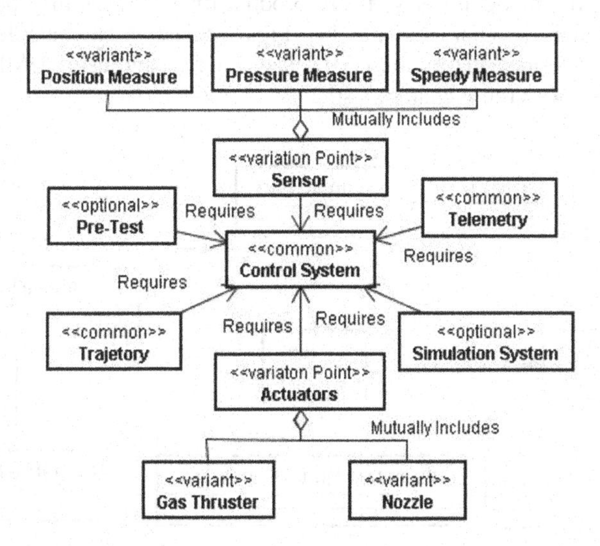

Fig. 3. Example of Feature Model for the BSL

3.2 Variability Analysis and Categorization

After construction of feature model it is possible to start the variability management activity. Van Gurp [17] mentions in his work that the first step to accomplish variability management is the variability identification activity.

Authors like [3, 4, 14 ,17] present some manners of variability categorization. Based on such works, variants associated to a variation point must be analyzed by the developer of the product line so as to be expressed, in general, as:

- *Mandatory Alternative*: a variant which is mandatory;
- *Optional*: a variant which may be required or not;
- *Alternative Inclusive*: it is necessary to choose one or more variants;
- *Alternative Exclusive*: only one variant is required.
- *Alternative Mutually Inclusive*: two variants together are always required.

3.3 Variability Modeling

According to the variability categories described in the previous section 3.2 and taking as basis the variability model in product line presented in [2], it is possible to generate the variability generic model shown in Figure 4.

Figure 4 shows that a core asset is constituted by one common part and may be constituted by zero or more variation points. Each variation point may be constituted by zero or many variants. Variants may be mandatory, optional, alternative inclusive, alternative exclusive or mutually inclusive according to the classification presented in section 3.2.

Every variation point may have zero or more variation mechanisms. More details on the variation mechanism of the proposed strategy are presented in section 3.4.

Therefore, in the modeling of software product lines, classes may be categorized according to their reuse characteristics [4]. Figure 5 shows an example of the BSL variability model, represented in the UML class diagram and using UML stereotypes for distinction between reuse characteristics.

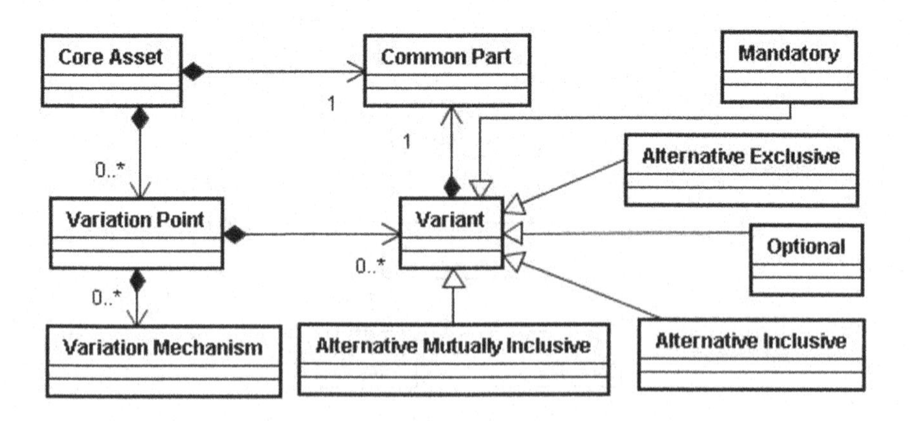

Fig. 4. Product Line Variability Model – Adapted from [2]

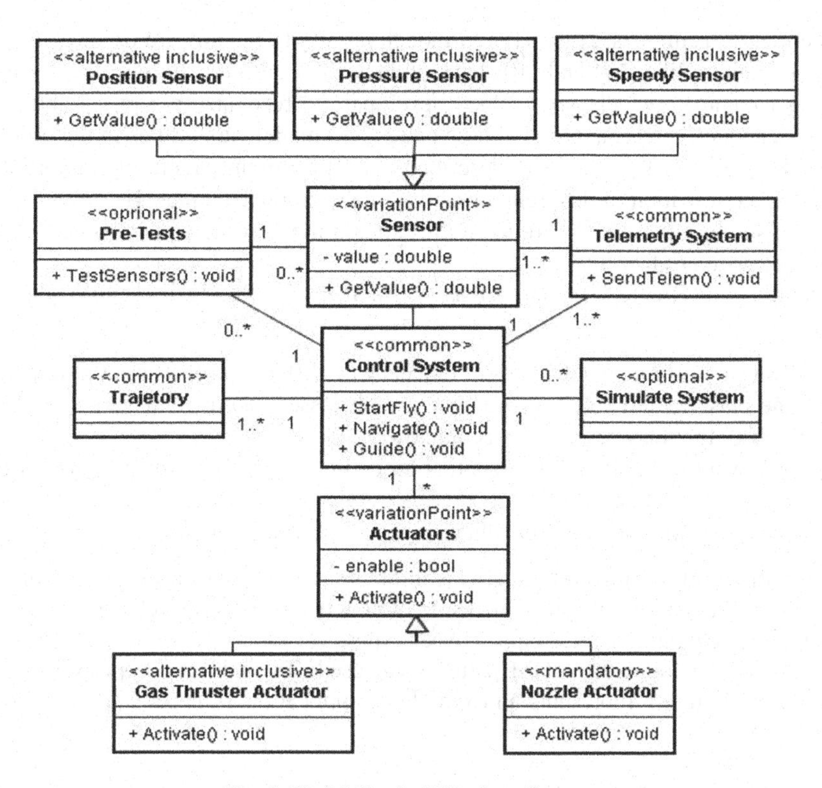

Fig. 5. Variability in BSL class diagram

3.4 Variation Mechanism

Bachmann [2] defines as variability mechanism the one that supports the creation and or selection of variants which are according to the restrictions for the core asset variation point. The author mentions suggestions for the variation mechanisms, such as: inheritance, parameterization, component substitution, templates, plug-ins, aspects and etc.

The work presented in [6] points out limitations to some variants creation techniques, such as:

- Parameterization: no functionality can change and select among alternative functionality is complicated and error prone, and is not recommended.
- Information Hiding: the application engineer chooses from a limited set of class implementations.
- Inheritance: although this approach allows the core assets developer to provide a large number of variants based on the parent class, the number of variants is still limited and the reuser must select a variant from a predefined set of choices.

One solution for such limitations, suggested by the proposed variability management strategy, is using a variation mechanism based on Adaptive Object Model (AOM) patterns.

The idea of using design patterns to model variability in software product lines is not new, authors like [17] and [10] have already used design pattern in their works, creating techniques which have helped developers to recognize where variability is needed. The work presented in [10] uses patterns to model variabilities in the product line in the design phase. The work described in [17] mentions pattens of variability to make the classification of different variability realization techniques. However, in the proposed strategy presented herein, AOM patterns are used as variation mechanism for the software product line.

According to Riehle [16] and Johnson [8], it is possible to use AOM patterns when:

- you want to create new types of objects with different properties at runtime;
- the class requires a lager number of subclasses and/or total variety of subclasses that may be required is unknown.
- you want to build applications that let end-users configure many types of object;
- the application require frequent changes and fast evolution;

The purpose of using AOM patterns is to obtain a model as generic as possible, as the set of patterns allows the easy inclusion of new types of objects, changing existing ones and changing the relationships between types.

To do so, the following AOM patterns are used: TypeObject, Property and Accountability. Figure 6 represents an example of application of TypeObject and Property patterns.

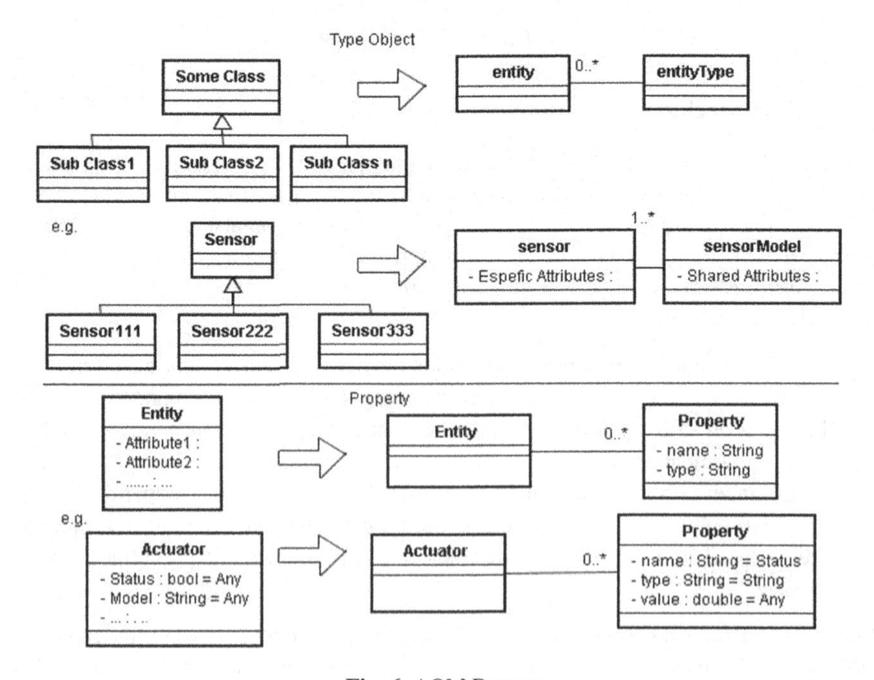

Fig. 6. AOM Patterns

TypeObject pattern separates an Entity from an Entity Type, makes an unknown number of the subclasses simple instances of a generic class. With TypeObject it is possible to create new instances of class types. The TypeObject pattern allows new classes to be created at runtime by means of instantiation of the generic class, theses new classes are instances called TypeObjects [8,18].

To vary the attributes of the new classes generated, the Property pattern is used, as through it each time a new attribute is required, a new class is created.

The Property pattern can define attributes contained in Entities, instead of each attribute being a different instance variable, the Property pattern makes an instance variable that holds a collection of attributes.

The application of these two patterns together results in an architecture called TypeSquare. Figure 7 represents the TypeSquare architecture.

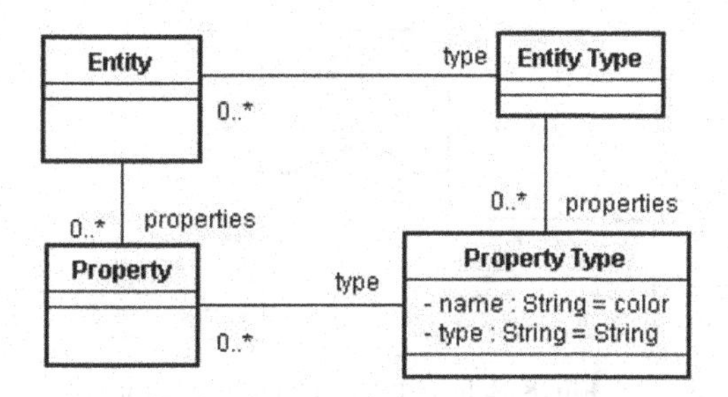

Fig. 7. TypeSquare [18]

Through the application of TypeSquare the system becomes able to define new high level Types of Properties.

The Accountability pattern, which manipulates types of relationships, is also used in the generic model. A more complete explanation on the application of AOM patterns may be found in [16, 18].

4 Variation Mechanism Based on AOM Applied to BSL

As an example, for the generation of new variants, AOM patterns were applied to sensors and actuators classes, part of the class diagram shown in Figure 5. Sensors and actuators classes were chosen because they are considered variation points and, therefore, have variants associated to these variation points. The purpose is to generalize the selected classes so that it is possible to create new instances of types of sensors and types of actuators, that is, new variants. Thus, AOM patterns are applied to highlighted classes in Figure 8.

The TypeObject pattern is used to generalize the type of Sensors and the types of Actuators. The Property pattern is used together with the TypeObject (TypeSquare) to define attributes and types of attributes of each type of Sensors and Actuators. In the same way, the Accountability pattern is used together with TypeObject to generalize the types of relationships between classes. In that way, it is possible to obtain a generic model of the BSL.

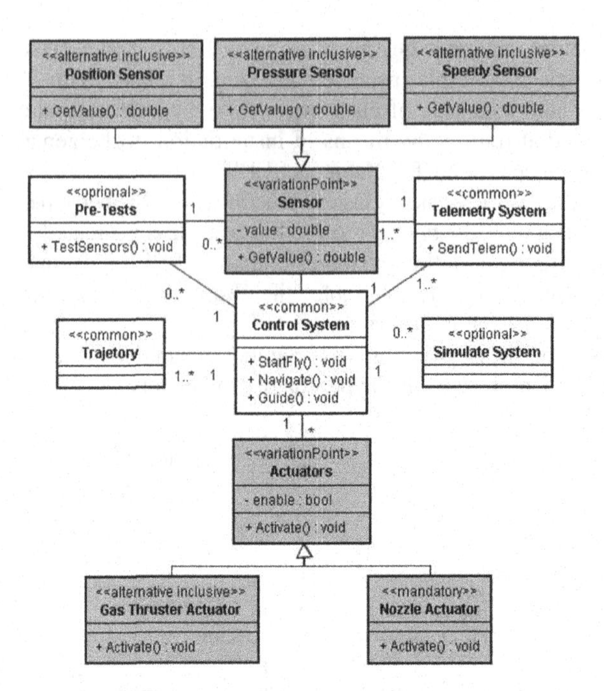

Fig. 8. Selected Sensors and Actuators

Figure 9 shows the resulting generic model after AOM patterns application to the classes related to sensors and actuators of BSL.

Thus, the model becomes generic enough to serve as base for the generation of other variants. It is now possible to manipulate different types of sensors and actuators with their respective attributes and relationships. Thus, from the generic model shown in Figure 9, it is possible to create a large number of new instances (variants) of a given class, without limitation of class numbers such as the one existing in the variability mechanism using information hiding and Inheritance.

Figure 10 shows created instances for classes Actuator, Actuator Type, Property and Property Type, based on Generic Model shown in Figure 9.

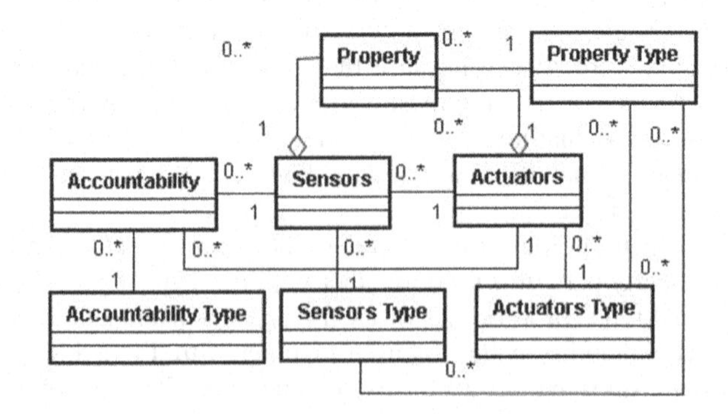

Fig. 9. Generic Model - BSL Sensors and Actuators

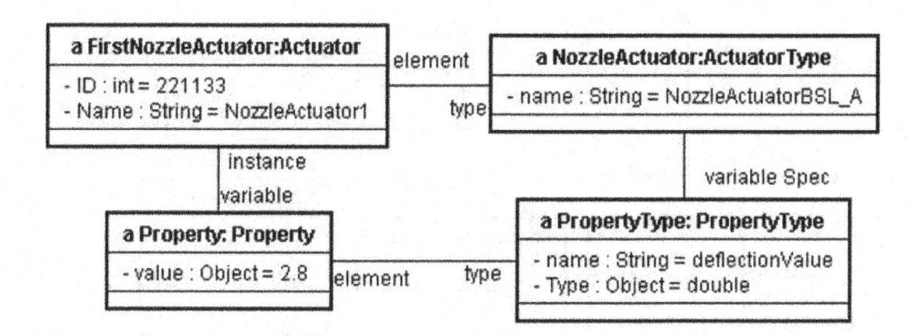

Fig. 10. Actuator Instance

To automate the variation mechanism the Variants Generator System (VGS) was created. The VGS is an application developed in Java language, having the function of supplying a java output file as a means of enabling the creation of UML graphic diagrams through proprietary tools such as Jude- UML Modeling Tool from Change Vision, Inc [9].

The application of design patterns enables the organization of information concerning the class, types of class, attributes, types of attribute, relationships and types of relationship in the database.

Figure11 shows the VGS interface, where the user may define for a given model, information on class, attributes and relationships types. The VGS developed uses such data in order to generate new variants.

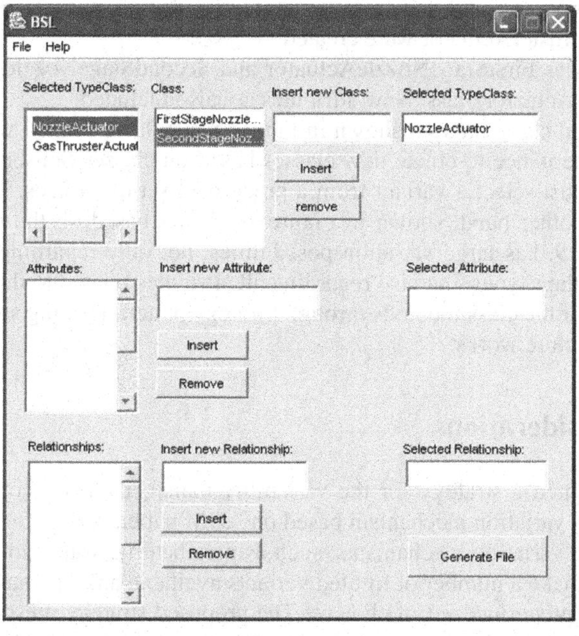

Fig. 11. VGS - Variation Generator System

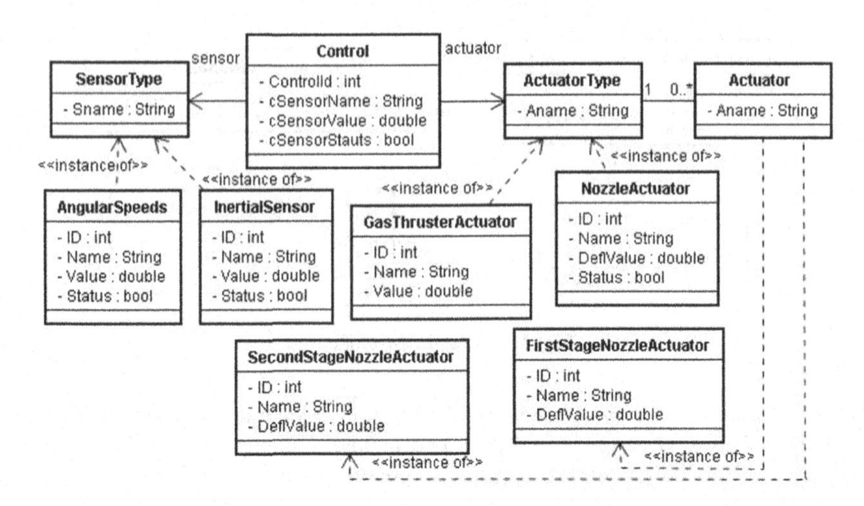

Fig. 12. Variants Sensors and Actuators

The VGS enables the inclusion or change of types of class, attributes or relationships, according to the user's selection and pre-established rules in the database. The established rules are based on the generated generic model.

Figure 12 shows part of a specific model of a BSL variation generated by the VGS, aided by a graphic tool.

Figure 12 shows that from the SensorType class, the class: InertialSensor and AngularSpeeds were created and from the ActuatorType class, the class: NozzleActuator and GasThrusterActuator were created.

Also, the classes FirstStageNozzleActuator and SecondStageNozzleActuator were created from the Actuator class. New attributes are also included.

The BSL initial class diagram shown in Figure 5 may have limited variation possibilities using inheritance to create new classes based on classes Sensor and Actuador and the reuser must select a variant from a predefined set of choices. The BSL class diagram, on the other hand, shown in Figure 12, being based on the generic model shown in Figure 9, has large variation possibilities, not only regarding the actuators and sensors and their types, but also regarding all attributes related to these classes.

The creation of the class methods through Strategy pattern is being studied and will be presented in future works.

5 Final Considerations

This work presented a strategy for the variability management for software product line which uses a variation mechanism based on AOM patterns.

Other existing variation mechanisms, such as parameterization, information hiding and inheritance make a number of limited variants available and the reuser must select a variant from a predefined set of choices. The proposed strategy presents an alternative solution to such limitations, using AOM patterns as variation mechanism to allow the core assets developer, in a system viewed as complicated, to provide a satisfactory number of variants. Thus, this variation mechanism based on AOM patterns provides more flexibility, as it allows changes in the system to be made more easily, through the use of a generic model. The variability strategy presented must be used when the product line is applied to the BLS. That is expected to improve the BLS development, increasing reuse, making information regarding the project institutionalized and the documentation updated.

References

[1] AEB- Agência Espacial Brasileira: Programa Nacional de Atividades Espaciais, Conselho Superior da Agência Espacial Brasileira. Brasília (2006), http://www.aeb.gov.br/conteudo.php?ida=2&idc=256 (accessed April 21, 2008)
[2] Bachmann, F., Clements, P.C.: Variability in Software Product Lines, Technical Report - CMU/SEI-2005-TR-012 (2005)
[3] Czarnecki, K., Eisenecker, U.W.: Generative Programming: Methods, Techniques, and Applications. Addison–Wesley, Indianapolis (2005)
[4] Gomaa, H.: Designing Software Product Lines with UML 2.0: From Use Cases to Pattern-Based Software Architectures. Addison-Wesley, Boston (2005)
[5] Gomaa, H., Shin, M.E.: Automated Software Product Line Engineering and Product Derivation. In: Proceedings of the 40th Annual Hawaii international Conference on System Sciences. IEEE Computer Society, Washington (2007)
[6] Gomaa, H., Webber, D.L.: Modeling Adaptive and Evolvable Software Product Lines Using the Variation Point Model. In: Proceedings of the 37th Annual Hawaii international Conference on System Sciences. IEEE Computer Society, Washington (2004)
[7] Griss, M.L., Favaro, J., Alessandro, M.: Integrating Feature Modeling with the RSEB. In: Proceedings of the 5th international Conference on Software Reuse – ICSR. IEEE Computer Society, Washington (1998)
[8] Johnson, R., Woolf, B.: Type Object. In: Robert, M., Dirk, R., Frank, B. (eds.) Pattern Languages of Program Design 3. Addison-Wesley, Reading (1997)
[9] Jude UML Modeling Tool (2008), http://jude.change-vision.com/jude-web/index.html (accessed April 19, 2008)
[10] Keepence, B., Mannion, M.: Using Patterns to Model Variability in Product Families. IEEE Software 16(4), 102–108 (1999)
[11] Kim, Y., Kim, J., Shin, S., Baik, D.: Managing Variability for Software Product-Line. In: Proceedings of the 4th International Conference on Software Engineering Research, Management and Applications, Washington (2006)
[12] Morais, P.: Programa de Veículos Lançadores de Satélites Cruzeiro do Sul - O Brasil na Conquista de sua Independência no Lançamento de Satélites. IAE CTA (2005), http://www.aeroespacial.org.br/aab/downloads.php (accessed April 21, 2008)
[13] Northrop, L.M., Clements, P.C.: A Framework for Software Product Line Practice, Version 5.0. Software Engineering Institute SEI (2008), http://www.sei.cmu.edu/productlines/framework.html (accessed April 21, 2008)
[14] Oliveira, E.A., Gimenes, I.M., Huzita, E.H., Maldonado, J.C.: A variability management process for software product lines. In: Proceedings of the Conference of the Centre for Advanced Studies on Collaborative Research, Ontario (2005)
[15] Pohl, K., Böckle, G., Linden, F.J.: Software Product Line Engineering: Foundations, Principles and Techniques. Springer, New York (2005)

[16] Riehle, D., Tilman, M., Johnson, R.: Dynamic Object Model. In: Proceedings of the 7th Conference on Patterns Languages of Programs - PloP 2000, Illinois (2000)
[17] Van Gurp, J., Bosch, J., Svahnberg, M.: On the Notion of Variability in Software Product Lines. In: Proceedings of the Working IEEE/IFIP Conference on Software Architecture. IEEE Computer Society, Washington (2001)
[18] Yoder, J.W., Balaguer, F., Johnson, R.: Architecture and Design of Adaptive Object-Models. ACM Sigplan Notices 36(fasc. 12), 50–60 (2001)

Use of Questionnaire-Based Appraisal to Improve the Software Acquisition Process in Small and Medium Enterprises

Ivan Garcia, Carla Pacheco, and Pavel Sumano

Postgraduate Department, Technological University of the Mixtec Region
Huajuapan de Len, Oaxaca (Mexico)
ivan@mixteco.utm.mx, leninca@mixteco.utm.mx, sumanoop@mixtli.utm.mx

Summary. This paper aims to show the application of a "Maturity Questionnaire" in a disciplined way. A Maturity Questionnaire typically is based on the Software Engineering Institute (SEI) published questionnaire; it represents a rigorous technique to collect data in a facilitated manner. The proposed questionnaire focuses in Supplier Agreement Management Process Area of the Capability Maturity Model Integration for Development v1.2. The objective is to obtain a snapshot of the Supplier Agreement Management Process (as a part of a process improvement program), to get a quick judgment of the capability level, and/or a precursor to a full assessment. It is expected that the application of the questionnaire will help to identify those practices that are performed but not documented, which practices need more attention and which are not implemented due to bad management or unawareness.

1 Introduction

In last years a great number of organizations have been interested in Software Process Improvement (SPI) [15]. An indicator of this interest is the increasing number of international initiatives related to SPI, such as CMMI-DEV [17], ISO/IEC 15504:2004 [9], SPICE [10], and ISO/IEC 12207:2004 [11].This approach has shown that a wide range of benefits can be expected by these organizations through adoption of software best practices [7]: software development making best use of currently available methods and technologies, in the most appropriate way according to business needs. More organizations are initiating SPI efforts. Once they are committed to begin the SPI effort, the second step is the assessment of the current capability of the organization process to develop and maintain software. Next, they have to define the action plans related to the implementation of the selected processes to be improved. A process assessment involves a disciplined examination of the processes within an organization. An assessment method requires: A measurement scale -most commonly a series of maturity or capability levels; criteria for evaluation against the scale, usually an underlying maturity model; a set of standards, best practices and industry norms; and a clear mechanism for representing the results. The assessment model provides a common framework for assessing processes, forms a basis for comparison of results from different assessments. Many assessment techniques or methods

R. Lee (Ed.): Soft. Eng. Research, Management & Applications, SCI 150, pp. 15–27, 2008.
springerlink.com

are available: CBA-IPI (Capability Maturity Model Based Appraisal for Internal Process Improvement) [7], Software Capability Evaluations, SCAMPI (Standard CMMI Appraisal Method for Process Improvement) [14], interview-based, document intensive, questionnaire-based, mini-assessment. The assessment is used because it checks process improvement progress (against some standard), gain insight into program/project process, check the maturity level, and serve as an audit tool -to check a contractor/supplier in a procurement/contractual situation. The Questionnaire-Based Appraisal (QBA) is an assessment tool that can be applied to many people, it is cost effective and non-invasive, provides quantitative data, and is possible to analyze the results with promptness [8]. Besides, it has been argued that the application of questionnaires consumes less time, effort and financial resources than other methods of data collection such as interviews and document reviews [1].

2 The CMMI-DEV Supplier Agreement Management Process

The purpose of CMMI-DEV Supplier Agreement Management process (CMMI-DEV SAM) is to manage the acquisition of products from suppliers for which there exists a formal agreement. This process area primarily applies to the acquisition of products and product components that are delivered to the project's customer. To minimize risks to the project, this process area may also be applied to the acquisition of significant products and product components not delivered to the project's customer (for example, development tools and test environments). Suppliers may take many forms depending on business needs, including in-house vendors (i.e., vendors that are in the same organization but are external to the project), fabrication capabilities and laboratories, and commercial vendors. The CMMI-DEV SAM recommended practices are shown in Table 1.

3 Data Collection Instruments: A Brief Review

This study has been defined by taking into account the generic SPI model defined by ISPI (Institute for Software Process Improvement Inc.) with four stages (commitment to appraisal, assessment, infrastructure and action plan, and implementation) (see Figure 2). Their objectives are similar to those of the IDEAL model [12] from the SEI. It must not be forgotten that this study focuses on phase 2 of the SPI Model: The Software Process Assessment. The IDEAL stages are:

Stage 1: Commitment to Improvement. Its objective is to obtain the support of senior management to carry out the improvement project.

Stage 2: Software Process Assessment. Its objective is to obtain the strengths and weaknesses of the process assessed with respect to a process model -CMMI-DEV. From this assessment, processes (usually 1 to 3) to be improved are selected.

Stage 3: Action Plans and Infrastructure to the Improvement. Its objective is to provide the needed infrastructure to carry out improvement (in selected

Table 1. CMMI DEV-SAM practices

Effective Practices	**Description**
1 Determining the type of acquisition	Determine the type of acquisition for every product or component to be acquiring (buy of products COTS, obtaining product across a contractual agreement or an external seller).
2 Select the supplier	Select suppliers being based on an evaluation of their skills to satisfy the specified requirements and the established criteria.
3 Establish agreements with the supplier	Establish and to maintain *formal agreements.*
4 Execute the supplier agreement	Develop the activities with the supplier according to the agreement.
5 Monitor selected supplier processes	Select, monitor, and analyze processes used by the supplier.
6 Evaluate selected supplier products	Select and evaluate products from the supplier of custom-made products.
7 Product acceptance	Ensure that the agreement has been satisfied before accepting the acquired product.
8 Product transition to the project	Assure that exist facilities adapted to receive, to store, to use and to support the acquired products, as well as the appropriate training.

processes) and to create the plans to follow on in order to define and implement improvements.

Stage 4: Implementation and Institutionalization. Its objectives are to define the new processes selected (following the previous plans) and to implement them in pilot projects. Finally, improvement must be institutionalized.

There is a wide number of data collection instruments that can be used for appraisals: questionnaires, surveys, interviews, and reviewing documentation, each having its own advantages and disadvantages. One of the commonly used techniques is a questionnaire. This is mainly because they can be applied to many people, they are cost effective, non-invasive, provide quantitative data, and results can be analyzed promptly [8]. However, it is important to mention that this technique lacks precision and also is easily open to misinterpretation. Questionnaires can be classified into *open and closed questions.* An open-question provides more information than a closed one. The complexity of analyzing data provided by open questions, however, is higher than those in closed-questions [18]. Moreover, a closed-question provides less information but its results can be more easily analyzed and are obtained faster than with the open one. Consequently, for this research a questionnaire was developed using closed questions as the main instrument for collecting appraisal data. In order to propose a new instrument for collecting appraisal data, a review was performed of the questionnaires available in the literature. The first questionnaire to be reviewed was the SEI Maturity Questionnaire [19]. The major disadvantage with this questionnaire is that it was developed for the SW-CMM model and cannot, therefore, be applied as it is to the CMMI-DEV SAM model. Furthermore, the maturity questionnaire provides little information about the CMMI-DEV SAM because it focuses on the maturity of the process without paying attention to find weaknesses in the practices. Another disadvantage is that this questionnaire is limited in the number of responses that can be selected: *Yes, No, Does not Apply* and *Do not Know*. In fact, there are only two options - Yes and No, because Does not Apply and Do not Know are used to validate the application of the questionnaire. Using the maturity questionnaire limits the information to two extreme ends: Yes, if the practice is performed and No if the practice is not performed. Therefore, it does not leave room for intermediate points. There are, for example, no options to pick up cases where practices are performed but rarely documented or when they are not documented at all. This type of question cannot be addressed with the options provided in the Maturity Questionnaire. Questionnaires with limited answer options may provide limited or misleading information. For example, a project sponsored by the SEI "CMMI Interpretive Guidance Project" supports this argument [5]. The questionnaire was applied to more than 600 people and the results report the following:

"We are not providing the results of the Generic Goals and Practices and Specific Process Areas sections of the Web-based questionnaire in this preliminary report. In both of these sections, there were no radio buttons and therefore the responses provided were in the form of specific comments. Many of these specific comments contain little information. For example, responses such as 'none' or 'no' were common" [5].

However, in one question of the same project, the SEI used five possible responses: *Almost always, More often than not, Sometimes, Rarely if ever* and *Do not know*. As a result, more distributions of the types of responses were obtained (see Figure 1). The report does not explain, however, the reasons why this methodology was not used in the same way for specific and generic practice questions. The report of the Process Improvement Program for the Northrop Grumman Information Technology Company [14] proposes a Questionnaire-Based Appraisal with seven possible responses: *Does Not Apply, Do not know, No, about 25% of the time, about 50% of the time, about 75% of the time*, and *Yes*. This work proposes more response granularity. It does not, however, explain how to apply this questionnaire to the Supplier Agreement Management process. Another disadvantage is that this report used the SA-CMM as a reference model and it focuses on the Software Acquisition process.

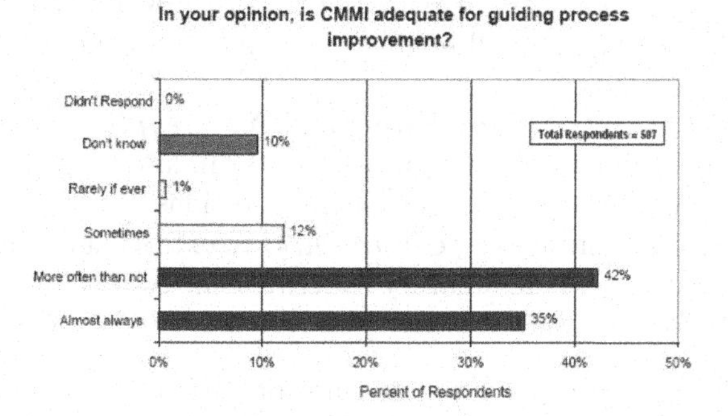

Fig. 1. Example of answer distribution

Another study reviewed was the software improvement model proposed by the ISPI. This model was used by [2], [3] and [4] in their research. For the appraisal stage, they proposed a questionnaire structure using five types of responses: ***Always*** when the practice is documented and performed between 100% - 75% of the time, ***More often*** when the practice is documented and performed between 74% - 50% of the time, ***Sometimes*** when the practice is not documented and is performed between 49% - 25% of the time. ***Rarely*** when the practice could be documented or not and is performed between 25% - 1 of the time. And ***Never*** when the practice is not performed in the organization. The response granularity is similar to that of Marciniak and Sadauskas [13] and provides more information about the current state of practices. This study only provides general information about the process without covering the CMMI-DEV SAM in full detail and without proposing precise actions for process improvement. Moreover, this questionnaire was designed for SW-CMM. The last study reviewed was the questionnaire proposed by Cuevas and Serrano [6]. This study proposes an assessment methodology based on a questionnaire to identify which practices of the Requirements Management process are performed but not documented, which practices require to be prioritized and which are not implemented due to bad management or unawareness. Cuevas's questionnaire is based on CMMI v1.1 [16] and only covers the requirements management process. In summary, the questionnaires reviewed here are deficient in their design and do not obtain relevant information. Furthermore, there is no evidence of a questionnaire that addresses the CMMI-DEV SAM in detail and there is no evidence of a questionnaire that covers both generic and specific practices.

4 The CMMI-DEV SAM Two-Phase Questionnaire

Based on the previously reviewed literature, a two-phase questionnaire is proposed. The questionnaire uses closed questions and limits the number of possible responses to seven. These are organized as follows:

- Five level-perform-answers: ***Always, Usually, Sometimes, Rarely if ever***, and ***Never.*** These will enable us to know the extent to which each practice is performed.
- Two validity-answers: ***Do not Know*** and ***Not Apply***. These will be used to appraise the validation of the questions, to validate the correctness of the question, and to check the syntaxes of the questions.
- Additional information spaces ***(Comments)*** to extract supplementary background information. It is mandatory to write some comments when checking any of the validity-answers.

Each possible response has a unique interpretation and indicates the performance level of a Supplier Agreement Management practice as described in Table 2.

The level-perform-answers determine the percentage in which each practice is performed. This varies from 'Never' with a value equal to 0, 'Rarely if ever' with a value equal to 1, 'Sometimes' with a value equal to 2, and 'Usually' with a value equal to 3, and 'Always' with a value equal to 4. The validity-answers do not have numerical value. Giving a specific weight to each response will enable us to easily

Table 2. Perform Level Classification

Possible Answer	Perform Level	Description
Always	4	The activity is documented and established in the organization. It is realized always, among 75 and 100 % in the organization software projects.
Usually	3	The activity is established in the organization but rarely documented. It is realized usually, among 50 and 75 % in the organization software projects.
Sometimes	2	The activity is weakly established in the organization. It is realized some times, among 25 and 50 % in the organization software projects.
Rarely if Ever	1	The activity is performed rarely in the organization. It is realized rarely, among 1 and 25 % in the organization software projects.
Never	0	The activity is not performed in the organization. It does not exist person and/or any group that perform the activity in the organization.
Do Not Know		The person is not sure of how to answer to the question.
Not Apply		The question does not apply for the organization.
Comments		This space is for elaborating or qualifying one's response to a question, and it is mandatory when one selects Do not know or Not Apply options.

First Phase – Supplier Agreement Management Practices

ESTABLISH SUPPLIER AGREEMENTS

I. Select Suppliers.
Select suppliers based on an evaluation of their ability to meet the specified requirements and established criteria.

Question	Always	Usually	Sometimes	Rarely if ever	Never	Don't know	Not Apply
1. Do you establish and document criteria for evaluating potential suppliers? *Consider using DAR when evaluating potential suppliers. This practice describes some of what is involved in such an evaluation.*	☐	☐	☐	☐	☐	☐	☐
Comments:							
2. Have you identified potential suppliers and distributed solicitation material and requirements to them? *A proactive manner of performing this activity is to conduct market research to identify potential sources of candidate products to be acquired, including candidates from suppliers of custom-made products and vendors.*	☐	☐	☐	☐	☐	☐	☐
Comments:							
3. Do you evaluate proposals according to evaluation criteria? *The proactive approach provides benefits such as addressing a capability gap of the organization uniformly, reducing time that projects take to select suppliers, establishing more efficient umbrella agreement with a preferred supplier, and protecting core competencies.*	☐	☐	☐	☐	☐	☐	☐
Comments:							
4. Do you evaluate risks associated with each proposed supplier? *Risks are typically included as criteria in a formal evaluation.*	☐	☐	☐	☐	☐	☐	☐
Comments:							
5. Do you evaluate proposed suppliers' ability to perform the work? *Examples of methods to evaluate the proposed supplier's ability to perform the work include the following:* • *Evaluation of prior experience in similar applications.* • *Evaluation of prior performance on similar work.* • *Evaluation of management capabilities.* • *Capability evaluations.* • *Evaluation of available facilities, resources, and staff to perform the work.*	☐	☐	☐	☐	☐	☐	☐

Fig. 2. An example of First Phase questions

analyze the results of the evaluation and to identify which practices are common within the whole organization and which ones are not performed at all.

The questionnaire proposed here has been based on the two types of practices established by the CMMI-DEV and is divided into two phases. The first phase is related to specific practices while the second-phase is related to generic practices. Another reason of this division is to differentiate the type of audience to whom it is applied. The first phase is aimed at employees who implement the process and is based on the specific practices from Supplier Agreement Management process of the CMMI-DEV [17]. This phase is divided into activities (showed in Table 1) that will be performed to achieve a well established Supplier Agreement Management process (see Figure 2).

The second phase is aimed at higher-level management such as general managers, system managers, software managers, or team leaders, and is based on the generic practices from the Supplier Agreement Management of the CMMI-DEV [17]. The application of this phase aims to find those activities for managing the software projects whether they are institutionalized or not and if they can support a managed process (see Figure 3). A managed process is a performed (Level 2) process that has the basic infrastructure in place to support the process. It is planned and implemented in accordance with policy; it employs skilled people who have adequate resources to produce controlled outputs; it involves relevant stakeholders; it is monitored, con-trolled, and reviewed; and it is evaluated for adherence to its process description.

Second Phase – Supplier Agreement Management Generic Practices

INSTITUTIONALIZE A MANAGED PROCESS

The organization' process is institutionalized as a managed process.

1. Does the organization have a written policy or guideline for planning the supplier agreement management process? *This policy establishes the organizational expectations for establishing, maintaining and satisfying supplier agreements.*	Always	Usually	Sometimes	Rarely if ever	Never	Don't know	Not Apply
Comments:							

2. Does the organization have a documented plan to perform the supplier agreement management process? *Portions of this plan for performing the supplier agreement management process can be part of (or referenced by) the project plan as described in the Project Planning process area. Often, however, some portions of the plan reside outside of the project with an independent group, such as contract management.*	Always	Usually	Sometimes	Rarely if ever	Never	Don't know	Not Apply
Comments:							

3. Do you provide the adequate resources to perform the supplier agreement management process? *Provide adequate resources for performing the supplier agreement management process, developing the work products, and providing the services of the process.*	Always	Usually	Sometimes	Rarely if ever	Never	Don't know	Not Apply

Fig. 3. An example of Second Phase questions

To determine if a Supplier Agreement Management practice is institutionalized, it is necessary to perform the following activities:

- Adhere to organizational policies.
- Track a documented project plan.
- Allocate adequate resources.
- Assign responsibility and authority.
- Train the affected people.
- Be placed under version control or configuration management.
- Be reviewed by the people affected.
- Measure the process.
- Ensure that the process complies with specified standards.
- Review the status with higher-level management.

The cross analysis of the responses of both questionnaires enable us to know those Supplier Agreement Management practices that have been covered by the software team and that have been spread throughout the organization as an institutionalized process. Similarly, this cross analysis helps us to identify other issues related to the combination of both phases of this questionnaire.

5 Case of Study

We designed a controlled experiment with 15 enterprises in which we focus our efforts in a semiformal version of the SCAMPI approach [14]. An assessment team and 5 project managers were chosen, besides, we separated the 15 organizations into 5 categories according with its profiles. These project managers are professional who know the enterprise's culture and the way that the development projects are conduced. The project managers had received the CMMI-DEV basic training. In a similar manner the project managers had answered a structured questionnaire related to the CMMI-DEV Capability Levels 1, 2 and 3. This questionnaire focuses solely on Supplier Agreement Management process issues. The

Table 3. Profiles of the participating organizations

Organization	Activiy	Size	Personnel Size
1	Development, installation and maintenance of Interactive Voice Response and his associated services	50-100	0-25
2	Hardware & software services and integral solutions	+500	250-500
3	Management products and technologies, storage, integration and efficient data change	+500	250-500
4	Informatics Solutions	+500	0-25
5	Development of informatics solutions for small enterprises	0-25	0-25

answers are agreed with the way that the enterprise works. The characteristics of the evaluated organizations are described in Table 3.

Once the questionnaires were completed we proceeded to identify the strong and weak points in order to show the current situation of the organizations according to the CMMI-DEV SAM process area, so that it is indicated where they must focus their efforts to raise the quality of the software that they acquire and support. Figure 4 shows the coverage of CMMI-DEV SAM process area for every organization category.

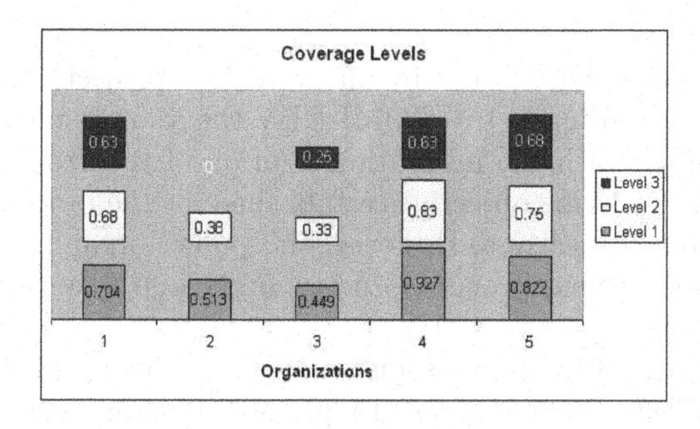

Fig. 4. Coverage Level of CMMI-DEV SAM Process per Levels

We determine that the process implemented in every organization has an acceptable average **(68%+38%+33%+83%+75% / 5=62%, over 50%)**. As a result of this study we observe that the process is a performed process that is also planned and executed in accordance with policy, have adequate resources to produce controlled outputs, but it does not involves relevant stakeholders; it is not monitored, controlled, and reviewed; and it is not evaluated for adherence to its process description.

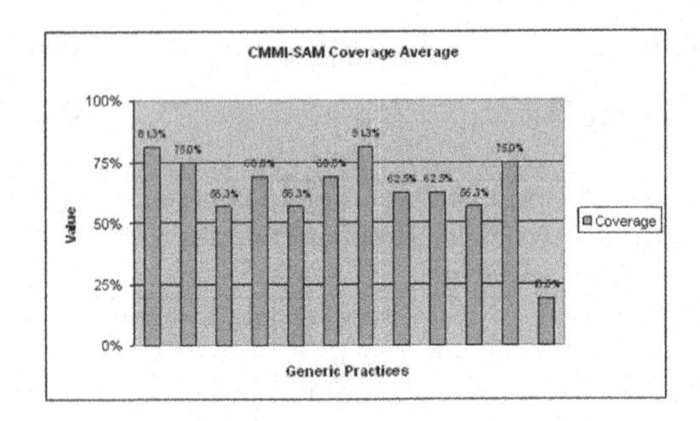

Fig. 5. Coverage average for the CMMI-DEV SAM generic practices

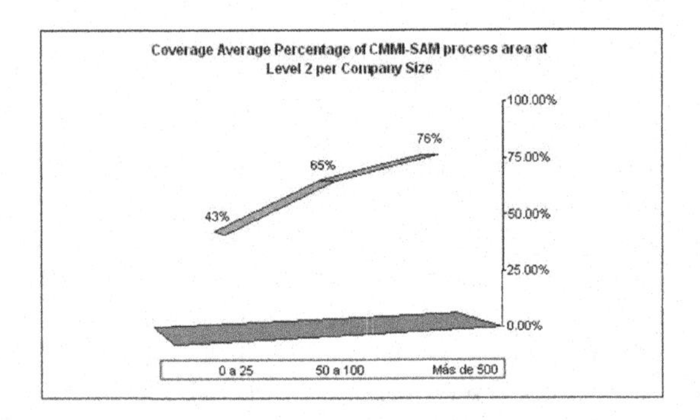

Fig. 6. Capability Average per Company Size

The process may be instantiated by an individual project, group, or organizational function. According to the CMMI-DEV the Management of the process (Level 2) is concerned with the institutionalization of the process area and the achievement of other specific objectives established for the process, such as cost, schedule, and quality objectives. However, the polled organizations fail in the same activities: documented process, product audit and take corrective actions as necessary.

Figure 5 shows the SAM-process institutionalization by practice, we could determine that the effective practices are: process planning, configuration management, verifying the process accomplishment and supervising with measure, while the non-effectives practices are: undocumented process, unskilled personnel (do not pay attention to personnel training), do not audit the work products and do not take corrective actions.

From Figure 6 we observe that the large organizations (with regard to its size) shown a defined process, focus their efforts to improve the acquisition process and use some recognized standard and procedures, while the small organizations shown that they are not prepared to assume the initiative of improve their acquisition process.

Besides the nonexistence of the practices mentioned previously, weaknesses were detected in the group of Software Quality Assurance because they do not verify the activities of quality assurance of the obtained software.

However, an increasing number of organizations show a special interest adopting the use of effective practices as an option to reduce the investment costs and implement new solutions, and increase their resource effectiveness by the adoption of CMMI-DEV SAM practices.

6 Conclusions

Though CMMI and ISO/IEC 15504 have exploded onto the market as models to follow when organizations try to apply process improvements, there are many organizations that are still not using these models. The CMMI is considered to be one of the best known models that focus on software process improvement for achieving quality software. The CMMI-DEV, however, is relatively new, so there is not much research written about which data collection instruments can be employed when using the CMMI-DEV approach. This research, therefore, developed an instrument to evaluate the current status of Supplier Agreement Management practices. The data collection instrument developed for the appraisal is a two-phase questionnaire. The evaluated organizations have acceptably developed their practices of Supplier Agreement Management process, though they should continue improving some aspects.

The large organizations need to manage many subcontractors. There is a possible cause that explains the greater coverage. The Small and Medium Enterprises (SMEs) are contracted by large enterprise and they do not need to deploy Supplier Agreement Management process. An important issue is the low coverage of institutionalization practices. There are necessary to support the improvement effort. QBA is a multidimensional, non-invasive technique and provides quantitative data. The use of questionnaires represents a low cost way to collect data. In the other hand, we proved that the QBA is a good technique (when it is used properly and in the right context) that can provide a good baseline for a new or existing process improvement program.

The questionnaire proposed here is divided into two phases. This division is mainly due to the fact that the CMMI-DEV clearly differentiates between specific practices and generic practices. As well as this, another reason for the division into two phases is because each section is applied to a different domain of people. In view of the foregoing, our future research efforts will focus on developing a methodology to implement the Supplier Agreement Management practices on SMEs internal processes. The two-phase questionnaire represents the first step in this research. The next step is related to more specific validation of the questionnaire. For this purpose, the questionnaire will be experimented on 26 SMEs through a project funded by the Mexican Ministry of Economics, Tourism and Trade. This research advocates, as future work, the idea of defining and implementing an "organizational repository of assets" where our questionnaires could be selected for any SMEs according to their needs.

Acknowledgement

This paper is sponsored by ENDESA, Everis Consulting Foundation and Sun Microsystems companies through "Research Group of Software Process Improvement in Latin America".

References

1. Brewerton, P.: Organizational research methods. SAGE, London (2001)
2. Calvo-Manzano, J.A., Cuevas, G., San Feliu, T., De-Amescua, A., Arcilla, M.M., Cerrada, J.A.: Lessons Learned in Software Process Improvement. The European Journal for the Informatics Professional IV(4), 26–29 (2003)
3. Calvo-Manzano, J.A., Cuevas, G., San-Feliu, T., De-Amescua, A., García, L., Pérez, M.: Experiences in the Application of Software Process Improvement in SMES. Software Quality Journal 10(3), 261–273 (2002)
4. Calvo-Manzano, J., Cuevas, G., San Feliu, T., Garcia, I., Serrano, A.: Requirements Management and Acquisition Management Experiences in Spanish Public Administrations. International Journal Information, Technologies and Knowledge 1(2) (2007) ISSN 1313-0455
5. CMMI Interpretive Guidance Project: Preliminary Report (CMU/SEI-2003-SR-007). Software Engineering Institute, Carnegie Mellon University, Pittsburgh, PA (October 2003)
6. Cuevas, G., Serrano, A., Serrano, A.: Assessment of the requirements management process using a two-stage questionnaire. In: Proceedings of Fourth International Conference on Quality Software 2004, QSIC 2004, September 8-9 (2004)
7. Dunaway, D.K., Masters, S.: CMM-Based Appraisal for Internal Process Improvement (CBA IPI): Method Description, Technical Report CMU/SEI-96-TR-007, Carnegie Mellon University, Software Engineering Institute, Pittsburgh (1996)
8. Gillham, B.: Developing a Questionnaire. Continuum, London, New York (2000)
9. ISO/IEC 155504-2:2003/Cor.1:2004(E). Information Technology - Process Assessment - Part 2: Performing an Assessment. International Organization for Standardization, Geneva (2004)
10. ISO/IEC TR 15504:1998(E). Information Technology - Software Process Assessments. Parts 1-9. International Organization for Standardization, Geneva (1998)
11. ISO/IEC 12207:2002/FDAM 2. Information Technology - Software Life Cycle Processes. International Organization for Standardization, Geneva (2004)
12. McFeeley, B.: IDEAL: A User's Guide for Software Process Improvement. Handbook; CMU/SEI-96-HB-001; Software Engineering Institute, Carnegie Mellon University (February 1996)
13. Marciniak, J.J., Sadauskas, T.: Use of Questionnaire-Based Appraisals in Process Improvement Programs. In: Proceedings of the Second Annual Conference on the Acquisition of Software-Intensive Systems, Arlington, Virginia, USA, p. 22 (2003)
14. Members of the Assessment Method Integrated Team. Standard CMMI® Appraisal Method for Process Improvement (SCAMPI), Version 1.1 (CMU/SEI-2001-HB-001). Software Engineering Institute, Carnegie Mellon University, December 2001, Pittsburgh, PA (2006)
15. Pino, F., García, F., Piattini, M.: Revisión Sistemática de Procesos Software en Micros, Pequeñas y Medianas Empresas. Revista Española de Innovación, Calidad e Ingeniería del Software 2(1), 6–23 (2006)

16. Software Engineering Institute. CMMI for Systems Engineering, Software Engineering, Integrated Product and Process Development, and Supplier Sourcing (CMMI-SE/SW/IPPD/SS, V1.1). Continuous Representation. CMU/SEI-2002-TR-011. Software Engineering Institute, Carnegie Mellon University
17. Software Engineering Institute. CMMI for Development (CMMI-DEV V1.2). CMU/SEI-2006 TR-008. Software Engineering Institute, Carnegie Mellon University
18. Yamanishi, K., Li, H.: Mining Open Answers in Questionnaire Data. IEEE Intelligent Systems 17(5), 58–63 (2002)
19. Zubrow, D., Hayes, W., Siegel, J., Goldenson, D.: Maturity Questionnaire (CMU/SEI-94-SR-7), Pittsburgh, PA, Software Engineering Institute. Carnegie Mellon University (June 1994)

Deploying Component–Based Applications: Tools and Techniques

Abbas Heydarnoori

School of Computer Science, University of Waterloo, Canada
aheydarnoori@uwaterloo.ca

Summary. Software deployment comprises activities for placing an already developed application into its operational environment and making it ready for use. For complex component-based applications that constitute many heterogeneous components with various hardware and software requirements, this deployment process can become one of the most burning challenges. In this situation, it is difficult to manually identify a valid deployment configuration that satisfies all constraints. Thus, automated tools and techniques are required to do the complex process of software deployment. To address this requirement, a variety of tools and techniques that support different activities of the deployment process have been introduced in both industry and academia. This paper aims to provide an overview of these tools and techniques.

Keywords: Software Components, Software Deployment, Deployment Life Cycle.

1 Introduction

Software deployment is a complex process that covers all post-development activities required to place an application into its target environment and make it available for use [1]. Along with significant advances in software development technologies in recent years, *component-based software development* (*CBSD*) has also gained a lot of attention in both industry and academia [2]. CBSD is a paradigm advancing a view of constructing software from reusable building blocks named *components*. According to Szyperski [3], a software component is a unit of composition with contractually specified interfaces and explicit context dependencies that can be deployed independently and is subject to composition by third parties. With respect to this definition, it is possible to have complex software systems that consist of a large number of heterogeneous components. In these applications, various components of the application often have different hardware and software requirements and hence they may provide their functionality only when those requirements are satisfied. Furthermore, software systems may undergo numerous updates during their lifetime and the components that comprise those systems are also more likely to be developed and released independently by third parties. Under these circumstances, finding a valid deployment configuration for a component-based application can be a challenging task and it might be effectively impossible to manually find a valid deployment configuration. This situation becomes even more complex in heterogeneous, resource-constrained, distributed, and mobile computing platforms

R. Lee (Ed.): Soft. Eng. Research, Management & Applications, SCI 150, pp. 29–42, 2008.
springerlink.com

that demand highly efficient software deployment configurations [4]. Thus, support for automated software deployment process becomes crucial.

To address this requirement, software deployment process has gained a lot of attention both in industry and research community in recent years and it is possible to find a large number of tools, technologies, techniques, and papers that address various aspects of the software deployment process from different perspectives. This paper aims to provide a survey of the existing software deployment tools and techniques in both industry and academia.

This paper is organized as follows. Section 2 considers some of the existing deployment processes in literature and proposes a generic deployment process. Section 3 provides a survey of software deployment tools in industry. Section 4 surveys the research-based techniques for the deployment of component-based applications. Finally, Section 5 concludes the paper.

2 Software Deployment Process

Software deployment is a sequence of related activities for placing a developed application into its target environment and making the application ready for use. Although this definition of software deployment is reasonable and clear, different sequences of activities have been mentioned in literature for this process [5, 6, 7, 8, 9, 10, 11]. This section has a look at some of the proposed deployment processes in literature and introduces a generic deployment process that covers the activities of all of them. This generic deployment process is then used in the rest of this paper to characterize different software deployment tools and techniques.

OMG Deployment and Configuration Specification (*OMG D&C Specification*) [5] outlines the following steps in the deployment process: *packaging* the application components; *installation* which involves populating a repository with the application components; *configuring* the functionality of the installed application in the repository; *planning* the deployment; *preparing* the target environment by moving the application components from the repository to the specified hosts; and *launching* the application.

Caspian [6] defines a five-step process for the deployment of component-based applications into distributed environments: *acquisition* of the application from its producer; *planning* where and how different components of the application should be installed in the target environment, resulting in a deployment plan; *installing* the application into its target environment according to its deployment plan; *configuring* it; and finally *executing* it.

Liu and Smith [7] in their LTS framework for component deployment specify the following activities for the deployment process: *shipping* the system from the development site; *installation* of the system at the deployment site; *reconfiguring* the system at the deployment site in response to changes; and actually *executing* the system.

The *Open Service Gateway initiative* (*OSGi*) [8] is an independent consortium of more than eighty companies launched in 1999 with the aim of developing a platform for the deployment of services over the wide-area networks to local networks and devices. In the deployment model of OSGi, a deployment process includes only the *install*, *update*, and *uninstall* activities.

Carzaniga et al. [9] in *Software Dock* research project define software deployment as an evolving collection of interrelated activities that are done on either the producer side or consumer side. The producer side activities include *release* and *retire*; the consumer side activities constitute *install*, *activate* (*launch*), *deactivate*, *reconfigure*, *adapt*, *update*, and *uninstall* (*remove*). Although this is one of the most comprehensive definitions of software deployment in literature, it lacks two activities in our view: *acquire* and *plan*. Acquisition is actually performed as part of the Software Dock's *install* activity and it is not explicitly introduced as a separate task. However, in many cases, the software systems are first acquired by the software consumers and then they are deployed on their operational environments. Moreover, as mentioned earlier, for large, distributed, component-based applications with many constraints and requirements, it is first required to plan how to install the components of the software system into its operational environment and then perform the actual installation. Hence, it is required to explicitly have a planning stage during the software deployment process. Finally, in the Software Dock deployment process, both of the *reconfigure* and *adapt* activities change the configuration of the deployed software system. The *reconfigure* activity assumes that it starts from a valid configuration of a system and then transforms it into another valid configuration. The *adapt* activity assumes it starts from an invalid configuration and then transforms it to a valid configuration. However, since both of these activities perform the same task which is changing the configuration of a software system into a valid configuration, we propose the *configure* activity rather than both of the Software Dock's *adapt* and *reconfigure* activities.

With respect to the above discussion, we propose a generic deployment process with ten activities that covers the activities of all the deployment processes mentioned in this section (Table 1). The description of these activities are as follows: (1) *release:* packages, prepares, and advertises a system for deployment into operational environments; (2) *acquire:* during this activity, the components of the application are acquired from the software producer and are put in a repository; (3) *plan:* given the specifications of the component-based application, the target environment, and user-defined constraints regarding this deployment, this activity determines where and how different components of the application will be executed in the target environment, resulting in a deployment plan; (4) *install:* this activity uses the deployment plan generated in the previous activity to install the application into its operational environment; (5) *configure:* this activity changes the configuration of an already installed software system; (6) *activate:* actually launches the application; (7) *update:* modifies a previously installed system and deploys a new, previously unavailable configuration of that system or updates the components of the system with newer releases of those components; (8) *deactivate:* shuts down executing components of an activated system; (9) *uninstall:* completely removes the software system from its operational environment; and (10) *retire:* makes a system release unavailable.

Table 1 compares the activities of different deployment processes and shows how they are mapped to the activities of the generic deployment process. However, it should be noted that the particular practices and procedures being done in each activity highly depend on the characteristics of the software system being deployed, its operational

Table 1. Mapping the activities of existing deployment processes to the activities of the generic software deployment process

Generic Software Deployment Process	Deployment Processes in Literature				
	OMG D&C	*LTS*	*OSGi*	*Software Dock*	*Caspian*
Release	Packaging	Shipping	-	Release	-
Acquire	Installation	-	-	-	Acquiring
Plan	Planning	-	-	-	Planning
Install	Preparation	Installation	Install	Install	Installation
Configure	Configuration	Reconfiguration	-	Adapt & Reconfigure	Configuration
Activate	Launch	Execution	-	Activate	Execution
Update	-	-	Update	Update	-
Deactivate	-	-	-	Deactivate	-
Uninstall	-	-	Uninstall	Uninstall	-
Retire	-	-	-	Retire	-

environment, and on the requirements of software developers and users. Therefore, the generic deployment process presented in this section can be customized for specific deployment requirements. For instance, Hnetynka and Murphy in [10] adapt the OMG D&C Specification for the deployment of Java-based components into embedded systems such as mobile phones and PDAs. As another example, Coupaye and Estublier in [11] present a deployment process for the deployment of complex applications into large companies.

3 Software Deployment Technologies in Industry

A variety of technologies exist in industry to support various activities of the software deployment process. This section surveys these technologies and classifies them into six major groups. Table 2 represents these groups and their corresponding example tools. This table also characterizes these deployment technologies in terms of their support for the activities of the generic software deployment process. In this table, • represents complete support, ∘ indicates partial support, and no circle means no support.

3.1 User-Driven Installers

There are many programs that are used to install and uninstall software systems from a single machine. Examples include InstallShield [12], InstallAnywhere [13], and Setup Factory [14]. These tools are typically not more than a compression tool with a user-friendly interface (e.g., a wizard). In these tools, different files of software systems are compressed into self-installing archives and are delivered to users. Then, users themselves use these tools to uncompress those archives on their intended machines. Users can also use those tools to uninstall the software systems by undoing the changes they have made during the installation. Many installers may also support some sort of

Table 2. Comparison of different industry-based deployment technologies in terms of their support for the activities of the generic software deployment process

Deployment Technology	Example Tools	Generic Software Deployment Process									
		Release	*Acquire*	*Plan*	*Install*	*Configure*	*Activate*	*Update*	*Deactivate*	*Uninstall*	*Retire*
User-Driven Installers	InstallShield, InstallAnywhere, Setup Factory	●			●	○		○		●	
Package Managers	Linux RPM, Fedora yum, Debian `Dpkg`	●		○	●	○		●		●	
Web-based Deployment Tools	Java Web Start, Windows Update, Microsoft Click-Once	●	●		●	○	●	●			
Systems Management Tools	Microsoft SMS, IBM TME, Altiris			●	●	●	●	●	●	●	
Remote Sessions	Citrix, PowerTCP, SSh						●				
Publish/Subscribe Tools	TIBCO Rendezvous, IBM Gryphon, Sun JMS	●	●		●			●			

configuration by which users can add or remove some functionalities from the installed software systems.

There are a number of limitations associated with user-driven installers. First, they are targeted to a single machine and it is typically impossible to use them for distributed platforms. Also, users themselves have to administer their software systems. This can have several difficulties such as: it is error prone, it is impossible to always rely on users, it might be difficult to enforce it, and finally it is hard to monitor.

3.2 Package Managers

Modern operating systems often come with package managers to assist in installing, uninstalling, and updating software systems. Linux RPM [15], Fedora yum [16], and Debian `Dpkg` [17] are representatives of this category of tools. All these tools are based on the concept of *package* and a *repository* that keeps information about the state of all installed packages. Each package comprises an archive of files to be deployed along with some metadata describing the software package such as its version.

The main goal of all package managers is to install packages in such a way that the correct dependencies among them are preserved. Package managers offer several functionalities such as creating a package, installing/uninstalling/updating a package, verifying the integrity of an installed package, querying the repository, and checking dependencies among installed packages. However, they do not take into account the execution phase of the deployment process and the package's content may not be executable code at all. Another issue of package managers is that they are targeted to a single machine and do not support distributed systems or large scale deployments. In addition, they are also user-driven and may pose the problems mentioned for user-driven installers.

3.3 Web-Based Deployment Tools

Web-based deployment tools try to use the connectivity and popularity characteristics of the Internet during the software deployment process. In these tools, it is not required to install or update the software system on every single host separately. Instead, the software application is deployed only to a single Web server. Then, client machines (users) connect to this server to download the application files or updates automatically. Representatives of these tools are Java Web Start [18], Microsoft Windows Update [19], and Microsoft ClickOnce [20]. However, one of the major limitations of these tools is that they are useless when there is no Internet connectivity.

3.4 Systems Management Tools

The term *systems management* is typically used to describe a set of capabilities (e.g., tools, procedures, policies, etc.) that enable organizations to more easily support their hardware and software resources. Systems management tools usually have a centralized architecture. In these tools, the IT administrator performs operations from a centralized location which is applied automatically to many systems in the organization. Therefore, the IT administrator is able to deploy, configure, manage, and maintain a large number of hardware and software systems from his own computer. Examples of these tools are Microsoft Systems Management Server [21], IBM Tivoli Management Environment [22], and Altiris Deployment Solution [23]. Systems management tools are all based on centralized repositories that keep deployment metadata such as client configurations and software packages. Moreover, they all support an inventory of hardware and software resources.

Systems management tools support many of the software deployment activities. In addition, all of the supported deployment activities can be done in a distributed environment as well. However, it is obvious that these tools are suitable for medium to large organizations. The issues associated with all these tools are that they are often heavy and complicated systems, they all require reliable networks, they are all based on complete administration control, and they are not viable for mobile devices.

3.5 Remote Sessions

Citrix [24], PowerTCP [25], and SSh [26] fall in this category of deployment tools. In this category, software systems are deployed to a single server machine. Then, each client initiates a session on that server and invokes desired software systems on it. Application state is kept entirely on the server. Therefore, these tools only support the execution activity of the deployment process.

The advantages of these tools are that they reduce the inconsistencies in deployed clients when the functionality is extended and it is not required to deploy the same application to several machines. The disadvantages are server load, under-utilized client resources, and consumption of network bandwidth.

3.6 Publish/Subscribe Tools

TIBCO Rendezvous [27], IBM Gryphon [28], and Sun Java Message Service [29] are examples of this kind of tools. In this class of tools, users express their interests ("subscribe") in certain kinds of events on a server, such as installing a new application or updating installed applications. Then, whenever these events happen on that server, they will be applied ("publish") automatically to the subscribed machines. This method is an efficient approach for the distribution of data from a source machine to a large number of target machines over a network.

The limitation of this class of tools is that the users themselves have to subscribe for the deployment of applications. Furthermore, these tools might not be efficient in costly and low-bandwidth networks.

4 Software Deployment Techniques in Research

Section 3 provided an overview of software deployment technologies in industry. However, The deployment of component-based applications has been the subject of extensive research in recent years. This section considers some of the deployment techniques proposed in the research community and classifies them into eight major deployment approaches. Table 3 represents these deployment approaches and their corresponding example techniques. This table further compares these deployment techniques in terms of their support for the activities of the generic software deployment process.

Table 3. Comparison of different research-based deployment approaches in terms of their support for the activities of the generic software deployment process

Deployment Approach	**Example Techniques**	**Generic Software Deployment Process**									
		Release	*Acquire*	*Plan*	*Install*	*Configure*	*Activate*	*Update*	*Deactivate*	*Uninstall*	*Retire*
QoS-Aware Deployment	DeSi			•							
	MAL			•	○	○		○			
	Caspian			•							
Architecture-Driven Deployment	Prism-DE		○	•	•		•	•			
	Olan	•	•		•	•	•				
Model-Driven Deployment	OMG D&C		•	•	•	•	•	○			
	Deployment Factory		•	•	•	•	•	○			
	DAnCE		•	•	•	•	•	○	•	○	
Agent-based Deployment	Software Dock	•	•		•	•		•			
	TACOMA	•	•		•			•			
Grid Deployment	Globus Toolkit			•	•	•	•				
	ORYA		•	○	•					•	
Hot Deployment	OpenRec		•	○	•	•	•	•	•	•	
	MagicBeans		•	○	•	•	•	•	•	•	
AI Planning-based Deployment	Sekitei			•							
	CANS		•	•	•	○		○		○	
Formal Frameworks	LTS	•	•	○	•	○	•	•		•	
	Conceptual				•						

However, this classification is not necessarily complete and it could be extended in the future. Furthermore, these deployment approaches are not completely disjoint and the same deployment technique might fall in two or more different deployment approaches. For instance, DAnCE [30] is a deployment technique that is both QoS-aware and model-driven. However, this categorization represents different directions of interest in research-based deployment techniques.

4.1 QoS-Aware Deployment

It is typically possible to deploy and configure a large and complex component-based application into its target environment in many different ways, specifically when the target environment is a distributed environment. Obviously, some of these deployment configurations are better than others in terms of some QoS (Quality of Service) attributes such as efficiency, availability, reliability, and fault tolerance. Thus, the deployment configuration has significant impacts on the system behavior and it is necessary to consider the issues related to the quality of a deployment. To this aim, a number of research approaches have been proposed in literature that address this aspect of software deployment. Examples include *DeSi* [31], *MAL* [32], and *Caspian* [6]. DeSi provides a deployment technique for maximizing the *availability* of systems defined as the ratio of the number of successfully completed inter-component interactions in the system to the total number of attempted interactions over a period of time. MAL is a system that enables the deployment of QoS-aware applications into ubiquitous environments. Ubiquitous systems such as Internet are those that can be instantiated and accessed anytime, anywhere, and by using any computing devices. Caspian provides a deployment planning approach in which the application being deployed and its target operational environment are modeled by graphs. Then, deployment is defined as the mapping of the application graph to the target environment graph in such a way that the desired QoS parameter is optimized. This deployment problem is solved for the QoS parameters *maximum reliability* and *minimum cost* in [33] and [34] respectively.

4.2 Architecture-Driven Deployment

The software architecture research community has also addressed configuration and deployment issues for component-based applications ([4], [35], [36], [37], [38]). A software architecture represents high-level abstractions for structure, behavior, and key properties of a software system. Software architectures are usually specified in terms of *components* that define computational units, *connectors* that define types of interactions among components, and the *configuration* that describes the topologies of components and connectors. For this purpose, *Architecture Description Languages* or *ADLs* have been developed to describe software systems in terms of their architectural elements.

Prism-DE [4] and *Olan* [35] are examples of deployment techniques that employ the concepts of software architecture during the process of software deployment. In these techniques, ADLs are used to specify valid deployment configurations. Then, during the process of deployment, candidate deployment configurations are checked against those valid deployment configurations.

4.3 Model-Driven Deployment

Model-Driven Architecture (*MDA*) [39] proposed by OMG is an approach for software development based on models in which systems are built via transformations of models. MDA defines two levels of models: *Platform-Independent Model* (*PIM*) and *Platform-Specific Model* (*PSM*). Developers begin with creating a PIM, then they transform the model step by step to a more platform-specific model until the desired level of specificity is approached. In the case of software deployment, the MDA approach starts with a platform-independent model of the target environment and the transformations finish with specific deployment configurations for the considered component-based applications [40]. Model-driven deployment has gained a lot of attention in recent years ([5], [30], [40], [41], [42], [43]). In particular, *OMG D&C Specification* [5] follows MDA concepts to provide a unified technique for the deployment of component-based applications into distributed environments. The OMG D&C Specification defines three platform-independent models: the *component model*, the *target model*, and the *execution model*. To use these platform-independent models with a specific component model, they have to be transformed to platform-dependent models, capturing the specifics of the concrete platform [44]. An example of this transformation to the *CORBA Component Model* (*CCM*) can be found in [5]. However, Hnetynka in [40] mentions that OMG's approach fails in building a single environment for the unified deployment of component-based applications and it leads to several deployment environments for different component models. To address this issue, he examines several common component models (e.g., *COM*, *CCM*, *EJB*, *SOFA*, and *Fractal*) and ADLs (e.g., *Wright* and *ACME*) to identify the set of features that are missing in the OMG's D&C component model. Based on this study, he proposes a unified deployment component model and introduces *Deployment Factory* (*DF*) as a model-driven unified deployment environment. As another example of model-driven deployment approaches, *DAnCE* [30] is a QoS-enabled middleware that conforms to the OMG D&C Specification. DAnCE enables application deployers to deploy component assemblies of distributed real-time and embedded (DRE) systems.

4.4 Agent-Based Deployment

A *mobile agent* is defined as an object that migrates through many hosts in a heterogeneous network, under its own control, to perform tasks using resources of those hosts [45]. A *Mobile Agent System* (*MAS*) is defined as a computational framework that implements the mobile agent paradigm. This framework provides services and primitives that help the implementation, communication, and migration of software agents. *Software Dock* [46] and *TACOMA* [47] are two MAS examples that use mobile agents for the purpose of software deployment. Software Dock is a deployment framework that supports cooperations among software producers themselves and between software producers and software consumers. To perform the software deployment activities, it employs mobile agents that traverse between software producers and consumers. Similarly, TACOMA is a deployment framework based on mobile agents for installing and updating software components in a distributed environment such as Internet.

4.5 Grid Deployment

A *computational grid* is defined as a set of efficient computing resources that are managed by a middleware that gives transparent access to resources wherever they are located on the network [48]. A computational grid can include many heterogeneous resources (e.g., computational nodes with various architectures and operating systems), networks with different performance properties, storage resources of different sizes, and so on. To take advantage of the computational power of grids, the application deployment must be as automated as possible while taking into account application constraints (e.g., CPU, Memory, etc.) and/or user constraints to prevent the user from directly dealing with a large number of hosts and their heterogeneity within a grid. Therefore, deployment of component-based applications into computational grids has been the subject of extensive research ([48], [49], [50], [51]).

4.6 Hot Deployment

In autonomic environments, a software system can automatically adapt its runtime behavior with respect to the configuration of the drastically changing execution environment and user requirements [52]. In this context, it is required to dynamically install, update, configure, uninstall, and replace software components without affecting the reliable behavior of the system or other constituent components. In doing so, a number of research projects address the ability of dynamically deploying software components during the program runtime, referred to as the *hot deployment* [7]. *OpenRec* [53] is a software architecture effort focusing on how to design dynamically reconfigurable systems. The *MagicBeans* platform [54] supports self-assembling systems of plug-in components that allow applications to be constructed and reconfigured dynamically at runtime. Reference [55] presents a generic architecture for the design and deployment of self-managing and self-configuring components in autonomic environments. Some other work on dynamic reconfiguration and hot deployment are [56], [57], [58], [59], [60], and [61].

4.7 AI Planning-Based Deployment

Planning the deployment of component-based applications into network resources has also gained attention in the AI research community. Two reasons have been mentioned for this [62]: (1) the requirement to satisfy the qualitative (e.g., reliability) and quantitative (e.g., disk quota) constraints; and (2) the fact that software deployment may involve selecting among compatible components as well as insertion of auxiliary components. Because of these reasons, AI planning-based techniques have been introduced for the purpose of software deployment. *Sekitei* [63] provides AI planning techniques for deploying components into resource-constrained distributed networks. *CANS* planner [64] finds optimal deployment of components along network paths. *Pegasus* [65] is a planning architecture for building grid applications. *Ninja* planner [66] makes directed acyclic graph (DAG) structured applications by using the available components in the network. *Panda* [67] has a database of predefined plan templates and simply instantiates a suitable template based on the programmer-provided rules that decide whether or not a component can be instantiated on a network resource.

4.8 Formal Deployment Frameworks

There are few works that provide platform-independent formal frameworks for the deployment of component-based applications. In these frameworks, different activities of the software deployment process are defined formally in a platform-independent manner that are suitable for derivation of theoretical results. For example, they can give deployment tool developers a theoretical basis to implement systems with well-defined behavior. Examples of these frameworks are *LTS* [7] and *conceptual foundation* [68]. LTS provides formalisms for almost all the activities of the software deployment process. The conceptual foundation only proposes conditions under which various software installation strategies are safe and successful.

5 Conclusions

An application can provide its expected functionality only when it is deployed and configured correctly in its operational environment. As a result, software deployment is a critical task that takes place after the development of an application. Inspired by the work by Carzaniga et al. [9], this paper defined software deployment as a process comprising ten activities related to releasing, acquiring, planning, installing, configuring, activating, updating, deactivating, uninstalling, and retiring software systems. However, for many modern component-based applications, the deployment process is complicated by the dimensions along which a system can be configured. Thus, a variety of tools and techniques have been introduced in both industry and academia to address this problem. This paper surveyed a set of representatives of these tools and techniques, and assessed them in terms of their support for the activities of the deployment process. This assessment indicated that there is no deployment tool or technique that can support the full range of deployment process activities. This suggests the deployment of component-based applications as an open problem that requires further research by the research community.

References

1. Heimbigner, D., Hall, R.S., Wolf, A.L.: A framework for analyzing configurations of deployable software systems. In: ICECCS (1999)
2. Crnkovic, I., Hnich, B., Jonsson, T., Kiziltan, Z.: Specification, implementation, and deployment of components. Commun. ACM 45(10), 35–40 (2002)
3. Szyperski, C.: Component Software - Beyond Object-Oriented Programming. Addison-Wesley, Reading (1999)
4. Mikic-Rakic, M., Medvidovic, N.: Architecture-level support for software component deployment in resource constrained environments. In: Bishop, J.M. (ed.) CD 2002. LNCS, vol. 2370. Springer, Heidelberg (2002)
5. Deployment and configuration of component-based distributed applications specification, `http://www.omg.org/docs/ptc/04-05-15.pdf`
6. Heydarnoori, A.: Caspian: A QoS-aware deployment approach for channel-based component-based applications. Technical Report CS-2006-39, David R. Cheriton School of Computer Science, University of Waterloo (2006)

7. Liu, Y.D., Smith, S.F.: A formal framework for component deployment. In: OOPSLA (2006)
8. OSGi Alliance, http://www.osgi.org/
9. Carzaniga, A., Fuggetta, A., Hall, R.S., Hoek, A.V.D., Heimbigner, D., Wolf, A.L.: A characterization framework for software deployment technologies, Technical Report CU-CS-857-98, Dept. of Computer Science, University of Colorado (1998)
10. Hnetynka, P., Murphy, J.: Deployment of Java-based components in embedded environment. In: IADIS Applied Computing (2007)
11. Coupaye, T., Estublier, J.: Foundations of enterprise software deployment. In: CSMR (2000)
12. InstallShield Developer, http://www.installshield.com/isd/
13. Zero G Software deployment and lifecycle management solutions, http://www.zerog.com/
14. Setup factory, http://www.indigorose.com/suf/
15. RPM package manager, http://www.rpm.org/
16. Yum: Yellow dog updater, http://linux.duke.edu/projects/yum/
17. Package maintenance system for Debian, http://packages.debian.org/dpkg/
18. Java web start technology, http://java.sun.com/products/javawebstart
19. Microsoft windows update, http://update.microsoft.com
20. ClickOnce: Deploy and update your smart client projects using a central server, http://msdn.microsoft.com/msdnmag/issues/04/05/clickonce/
21. Systems management server home, http://www.microsoft.com/smserver/
22. IBM Tivoli software, http://www.tivoli.com/
23. Altiris deployment solution, http://www.altiris.com/
24. Citrix, http://www.citrix.com/
25. PowerTCP, http://www.dart.com/powertcp/
26. Secure shell (SSH), http://www.ssh.com/
27. TIBCO Rendezvous, http://www.tibco.com/software/messaging/
28. IBM Gryphon, http://www.research.ibm.com/distributedmessaging/
29. Java message service (JMS), http://java.sun.com/products/jms/
30. Deng, G., Balasubramanian, J., Otte, W., Schmidt, D., Gokhale, A.: DAnCE: A QoS-enabled component deployment and configuration engine. In: Dearle, A., Eisenbach, S. (eds.) CD 2005. LNCS, vol. 3798. Springer, Heidelberg (2005)
31. Mikic-Rakic, M., Malek, S., Medvidovic, N.: Improving availability in large, distributed component-based systems via redeployment. In: Dearle, A., Eisenbach, S. (eds.) CD 2005. LNCS, vol. 3798. Springer, Heidelberg (2005)
32. Wichadakul, D., Nahrstedt, K.: A translation system for enabling flexible and efficient deplyoment of QoS-aware applications in ubiquitous environments. In: Bishop, J.M. (ed.) CD 2002. LNCS, vol. 2370. Springer, Heidelberg (2002)
33. Heydarnoori, A., Mavaddat, F.: Reliable deployment of component-based applications into distributed environments. In: ITNG (2006)
34. Heydarnoori, A., Mavaddat, F., Arbab, F.: Deploying loosely coupled, component-based applications into distributed environments. In: ECBS (2006)
35. Balter, R., Bellissard, L., Boyer, F., Riveill, M., Vion-Dury, J.Y.: Architecturing and configuring distributed application with Olan. In: Middleware (1998)
36. Quema, V., Cecchet, E.: The role of software architecture in configuring middleware: The ScalAgent experience. In: Papatriantafilou, M., Hunel, P. (eds.) OPODIS 2003. LNCS, vol. 3144. Springer, Heidelberg (2004)
37. Quema, V., et al.: Asynchronous, Hierarchical, and scalable deployment of component-based applications. In: Emmerich, W., Wolf, A.L. (eds.) CD 2004. LNCS, vol. 3083. Springer, Heidelberg (2004)

38. Matevska-Meyer, J., Hasselbring, W., Reussner, R.H.: Software architecture description supporting component deployment and system runtime reconfiguration. In: WCOP (2004)
39. OMG model driven architecture, http://www.omg.org/mda/
40. Hnetynka, P.: A model-driven environment for component deployment. In: SERA (2005)
41. Hoffmann, A., Neubauer, B.: Deployment and configuration of distributed systems. In: Amyot, D., Williams, A.W. (eds.) SAM 2004. LNCS, vol. 3319. Springer, Heidelberg (2005)
42. Belkhatir, N., Cunin, P., Lestideau, V., Sali, H.: An OO framework for configuration of deployable large component based software products. In: OOPSLA ECOOSE Workshop (2001)
43. Jansen, S., Brinkkemper, S.: Modelling deployment using feature descriptions and state models for component-based software product families. In: Dearle, A., Eisenbach, S. (eds.) CD 2005. LNCS, vol. 3798. Springer, Heidelberg (2005)
44. Bulej, L., Bures, T.: Using connectors for deployment of heterogeneous applications in the context of OMG D&C Specification. In: INTEROP-ESA (2005)
45. Rus, D., Gray, R., Kotz, D.: Transportable information agents. In: AAMAS (1997)
46. Hall, R.S., Heimbigner, D., Wolf, A.L.: A cooperative approach to support software deployment using the software dock. In: ICSE (1999)
47. Sudmann, N.P., Johansen, D.: Software deployment using mobile agents. In: Bishop, J.M. (ed.) CD 2002. LNCS, vol. 2370. Springer, Heidelberg (2002)
48. Lacour, S., Perez, C., Priol, T.: A software architecture for automatic deployment of CORBA components using grid technologies. In: DECOR (2004a)
49. Lacour, S., Perez, C., Priol, T.: Deploying CORBA components on a computational grid: General principles and early experiments using the Globus Toolkit. In: Emmerich, W., Wolf, A.L. (eds.) CD 2004. LNCS, vol. 3083. Springer, Heidelberg (2004)
50. Lestideau, V., Belkhatir, N.: Providing highly automated and generic means for software deployment process. In: Oquendo, F. (ed.) EWSPT 2003. LNCS, vol. 2786. Springer, Heidelberg (2003)
51. Brebner, P., Emmerich, W.: Deployment of infrastructure and services in the open grid services architecture (OGSA). In: Dearle, A., Eisenbach, S. (eds.) CD 2005. LNCS, vol. 3798. Springer, Heidelberg (2005)
52. Murch, R.: Autonomic Computing. Prentice-Hall, Englewood Cliffs (2004)
53. Hillman, J., Warren, I.: An open framework for dynamic reconfiguration. In: ICSE (2004)
54. Chatley, R., Eisenbach, S., Magee, J.: Magicbeans: A platform for deploying plugin components. In: Emmerich, W., Wolf, A.L. (eds.) CD 2004. LNCS, vol. 3083. Springer, Heidelberg (2004)
55. Patouni, E., Alonistioti, N.: A framework for the deployment of self-managing and self-configuring components in autonomic environments. In: WoWMoM (2006)
56. Akkerman, A., Totok, A., Karamcheti, V.: Infrastructure for automatic dynamic deployment of J2EE applications in distributed environments. In: Dearle, A., Eisenbach, S. (eds.) CD 2005. LNCS, vol. 3798. Springer, Heidelberg (2005)
57. Cervantes, H., Hall, R.S.: Autonomous adaptation to dynamic availability using a service-oriented component model. In: ICSE (2004)
58. Hicks, M.W., Moore, J.T., Nettles, S.: Dynamic software updating. In: PLDI (2001)
59. Liu, H., Parashar, M.: A component-based programming framework for autonomic applications. In: ICAC (2004)
60. Mitchell, S.R.: Dynamic Configuration of Distributed Multimedia Components. PhD thesis, University of London (2000)
61. Paula, J., Almeida, A., Wegdam, M., Sinderen, M.V., Nieuwenhuis, L.: Transparent dynamic reconfiguration for CORBA. In: DOA (2001)

62. Kichkaylo, T., Karamcheti, V.: Optimal resource-aware deployment planning for component-based distributed applications. In: HPDC (2004)
63. Kichkaylo, V., Ivan, A., Karamcheti, V.: Constrained component deployment in wide-area networks using AI planning techniques. In: IPDPS (2003)
64. Fu, X., Karamcheti, V.: Planning for network-aware paths. In: Stefani, J.-B., Demeure, I., Hagimont, D. (eds.) DAIS 2003. LNCS, vol. 2893. Springer, Heidelberg (2003)
65. Blythe, J., Deelman, E., Gil, Y., Kesselman, C., Agarwal, A., Mehta, G., Vahi, K.: The role of planning in grid computing. In: ICAPS (2003)
66. Gribble, S., et al.: The Ninja architecture for robust internet-scale systems and services. Computer Networks 35(4), 473–497 (2001)
67. Reiher, P., Guy, R., Yavis, M., Rudenko, A.: Automated planning for open architectures. In: OpenArch (2000)
68. Parrish, A., Dixon, B., Cordes, D.: A conceptual foundation for component-based software deployment. The Journal of Systems and Software 57(3), 193–200 (2001)

Towards Improving End-to-End Performance of Distributed Real-Time and Embedded Systems Using Baseline Profiles

James H. Hill and Aniruddha Gokhale

Vanderbilt University
Nashville, TN, USA
{j.hill,a.gokhale}@vanderbilt.edu

Summary. Component-based distributed real-time and embedded (DRE) systems must be properly deployed and configured to realize an operational system that meets its functional and quality-of-service (QoS) needs. Different deployments and configurations often impact systemic QoS concerns, such as end-to-end response time. Traditional techniques for understanding systemic QoS rely on complex analytical and simulation models, however, such techniques provide performance assurance at design-time only. Moveover, they do not take into account the complete operating environment, which greatly influences systemic performance.

This paper presents a simple technique for searching for deployments that improve performance along a single dimension of QoS concern, such as response time. Our technique uses baseline profiles and experimental observations to recommend new deployments. The results indicate that we are able to use baseline profiles to converge towards solutions that improve systemic performance.

Keywords: baseline testing, component-based development, deployment and configuration, performance analysis.

1 Introduction

Challenges of developing component-based systems. Component-based technologies are raising the level of abstraction so software system developers of distributed real-time and embedded (DRE) systems can focus more on the application's "business-logic". Moreover, component-based technologies are separating many concerns, such as deployment (*i.e.*, placement of component on host) and configuration (*i.e.*, setting of component properties) [1], management of the application's lifecycle [2], and management of execution environment's quality-of-service (QoS) policies [2, 12], of the application. The result of separating the concerns is the ability to fully address each concern independently of the application's "business-logic" (*i.e.*, its implementation). The technologies that are making this possible include, but are not limited to: CORBA Component Model (CCM), J2EE, and Microsoft.NET.

Although component-based technologies are separating many concerns of DRE system development, realizing systemic (*system wide*) QoS properties, which is a key characteristic of DRE systems, is being pushed into the realized system's deployment and configuration (D&C) [1] solution space. For example, it is hard for DRE system

R. Lee (Ed.): Soft. Eng. Research, Management & Applications, SCI 150, pp. 43–57, 2008.
springerlink.com

developers to fully understand systemic QoS properties, such as end-to-end response time of system execution paths until the system has been properly deployed and configured in its target environment. Only then would a DRE system developer determine if different components designed to collaborate with each other perform better when located on different hosts compared to collocating them on the same host, which in turn might give rise to unforeseen software contentions, such as waiting for a thread to handle an event, or event/lock synchronization, or many known software design anti-patterns [13] in software performance engineering [8].

Outside of brute force trial and error, traditional techniques, such as complex analytical and mathematical models [8,5,15], can be used to locate candidate D&Cs that meet the required systemic QoS properties. Although such techniques can locate candidate D&Cs, they provide design-time assurance and do not take into account the complete operating environment. Moreover, using these techniques are beyond the knowledge domain of component-based system developers. Component-based DRE system developers, therefore, need simpler techniques that are intuitive and easier to use that can provide assistance in understanding systemic QoS properties in relation to the realized system's D&C.

Solution Approach → Use baseline profiles to understand the deployment solution space. Each individual component in a component-based DRE system has a baseline profile that captures its performance properties when deployed in isolation to other components. As components are collocated (*i.e.*, placed on the same host) with other components, their actual performance may deviate from its baseline performance profile due to unforeseen software contentions. Ideally, to improve systemic QoS properties for a component-based system, system developers should evaluate each individual component's performance against its baseline profile. This will allow them to understand if a component is performing as expected. Moreover, it will allow them to pinpoint possible problems and recommend new deployments that could possibly improve systemic QoS properties.

This paper describes our solution approach for understanding systemic QoS properties, such as end-to-end response time of system execution paths, in relation to the deployment solution space of component-based DRE systems. Our solution uses baseline profiles for each component, which are obtained by instrumenting and profiling the component in a controlled environment. We then execute a deployment of the system in the target environment and compare its performance against the baseline profile. Finally, based on the percentage error between the current test and the baseline profile, and the history of previous deployments, we recommend a new deployment for experimentation that will either improve/degrade system QoS properties. Our solution is designed so that developers not only converge towards a D&C that ensures systemic QoS properties, but allows system developers to get a better understanding of how different deployments influence systemic QoS properties.

Paper Organization. The remainder of this paper is organized as follows: Section 2 introduces a case study to highlight challenges of software contention; Section 3 discusses our technique for understanding the deployment solution space; Section 4 discusses results of applying our technique to our case study; Section 5 presents related work; and Section 6 provides concluding remarks and future research.

2 Manifestation of Software Contention: A Case Study Perspective

In this section we use a representative case study from the shipboard computing environment domain to highlight how different deployment and configurations can introduce different software contention problems that impact systemic QoS properties.

2.1 The SLICE Shipboard Computing Scenario

The SLICE scenario is a representative example of a shipboard computing environment application. We have used this example in prior work and multiple studies, such as the evaluation of system execution modeling tools [4] and the formal verification [3]. In this paper we use it to highlight software contention issues arising out of different deployment and configurations. We briefly reiterate its description here.

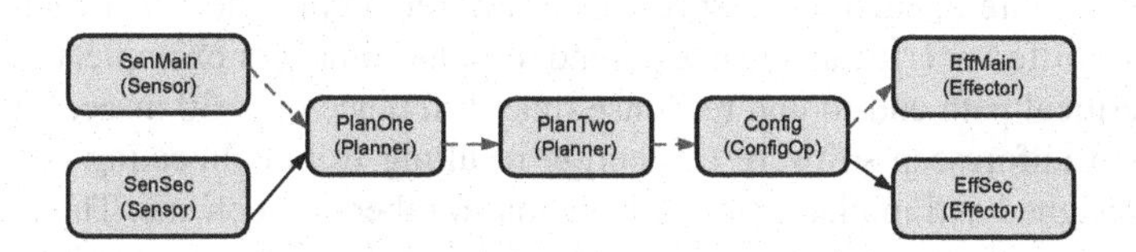

Fig. 1. High-level structural composition of the SLICE scenario

The SLICE scenario, which is illustrated in Figure 1, consists of 7 different component implementations, (*i.e.,* the rectangular objects): *SenMain*, *SenSec*, *PlanOne*, *PlanTwo*, *Config*, *EffMain*, and *EffSec*. The directed lines between the components represent connections for inter-component communication, such as input/output events. The components in the SLICE scenario are deployed across 3 computing nodes and `SenMain` and `EffMain` are deployed on separate nodes to reflect the placement of physical equipment in the production shipboard environment. Finally, events that propagate along the path marked by the dashed (red) lines in Figure 1 represent an execution (or critical) path that must have optimal performance.

2.2 Software Contention in the SLICE Scenario

We define software contentions as the deviation of performance as a result of the system's implementation—also characterized as performance anti-patterns [13]. In many cases hardware contention does not have noticeable affects on resource utilization. For example, it is possible to have high end-to-end response time and low CPU utilization if different portions of the software components (*e.g.*, threads of execution) unnecessarily hold mutexes that prevent other portions of the software (or components) from executing. The (software) component holding the mutex may not be executing on the CPU,

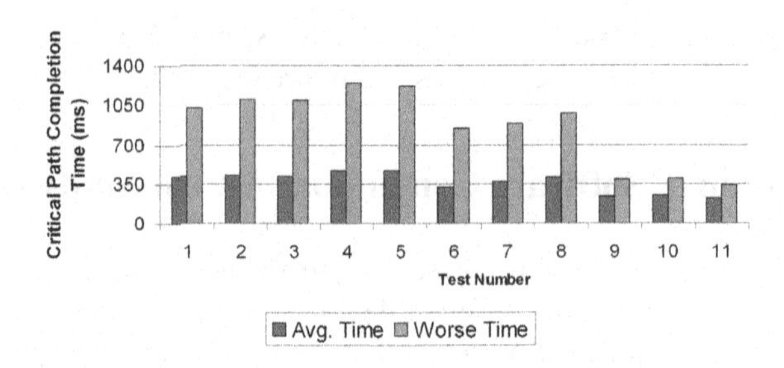

Fig. 2. Prior results of SLICE scenario

however, it can be involved in other activities, such as sending data across the network. Likewise, it is possible to have high response time if too many components are sending event requests to a single port of another component [10], which increases the queuing time for handling events.

Figure 2 illustrates the results from prior work [4] that used brute force ad-hoc techiques to improve critical path end-to-end response time for the SLICE scenario. As shown in Figure 2, each test respresents a different deployment and each deployment yields a different critical path end-to-end response time. For example, tests 10 and 11 have a critical path end-to-end response time difference of $\sim$ 30 msec. We believe this is due to unforeseen software contention resulting from collocating components that were designed and implemented in isolation to other components. The remainder of this paper, therefore, details how we use baseline profiles to understand the deployment solution space and improve systemic QoS performance due to unforeseen software contentions using the SLICE scenario as an example case study.

3 Using Baseline Profiles to Understand the Deployment Solution Space

This section discusses our technique for using baseline profiles to explore and understand the deployment solution space of component-based DRE systems.

3.1 Representing Deployment Solution Space as a Graph

After system composition, a component-based system becomes operational after its deployment and configuration in the target environment. There are, however, many ways to deploy a component-based systems, such as the SLICE scenario in Section 2. For example, Table 1 shows two unique deployments for the SLICE scenario where both differ by the placement of one component; the *SenMain* component on either Host1 or Host2.

When we consider all the unique deployments in the deployment solution space for a component-based systems, it is possible to represent such a space as a graph $G = (D, E)$ where:

- D a set of vertices $d \in D$ in graph G that represents a unique deployment d in the deployment solution space D. For example, deploying *SenMain* then *SenSec* on a

Table 1. Example of unique deployments in the SLICE scenario

Host	Deployment	
	A	B
1	SenMain, SenSec Config	SenSec, Config
2	EffSec, PlanOne	EffSec, PlanOne SenMain
3	PlanTwo, EffMain	PlanTwo, EffMain

single host H_i is the same as deploying *SenSec* then *SenMain* on the H_i. This is different from traditional bin packing, which is based on permutations [9].

- E is a set of edges $e \in E$ where $e_{i,j}$ is an edge between two unique deployments $i, j \in D$. Each edge e signifies moving a single component from one host to another host in the system.

Definition 1. *If $C_{h,i}$ is the set of components deployed on host h in system H for deployment i, then a unique deployment is defined as:*

$$\exists h \in H : C_{h,i} \neq C_{h,j}; i, j \in D \wedge i \neq j \tag{1}$$

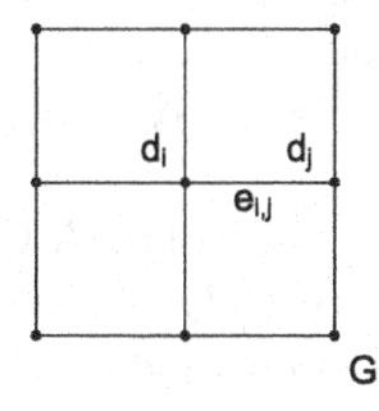

Fig. 3. Simple graph of component-based system deployment solution space

The resultant graph G, as shown in Figure 3, allows us to create a visual representation of the deployment solution space. The visualization, however, has no concrete meaning from the QoS perspective because developers cannot understand how one deployment differs from another deployment except for the unique combination of components on host. We address this problem by assigning a value to each edge e in graph G using Equation 2,

$$val(e) = \delta \tag{2}$$

which gives rise to a directed graph G'. Figure 4 illustrates G' where δ represents some difference of performance in a single QoS dimension, such as end-to-end response time or system throughput.

QoS, however, is known to comprise a N-dimensional space [11], and G' shown in Figure 4 is the deployment solution space for a single dimension of QoS, *i.e.*, visualized on a single plane. When we consider N dimensions of QoS, we create an N-planar

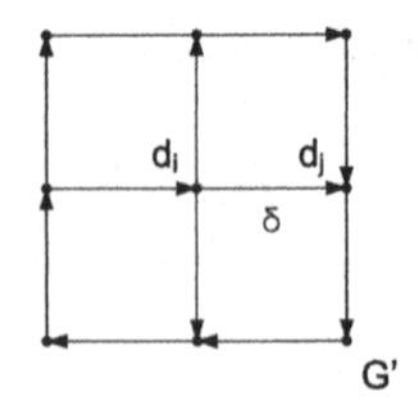

Fig. 4. Directed graph of component-based system deployment solution space

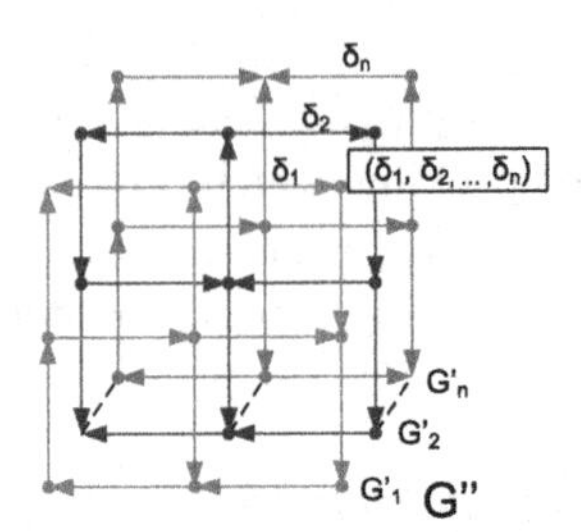

Fig. 5. N-planar directed graph of component-based system deployment solution space for N QoS dimensions

graph G''. Figure 5 illustrates G'' where each plane represents a single QoS dimension. The value of each edge between unique deployments in different planes, *i.e.*, QoS dimensions, forms a N-tuple vector (δ_1, δ_2, ..., δ_n) where $\delta_i = val(e \in G'_i)$ and $i \in$ QoS dimension, $G'_i \in G''$.

When also considering the N-planer graph of the N QoS dimension solution space, an edge between the same deployment for each QoS dimension is governed by Property 1. This results in a N-tuple vector equal to zero and is, therefore, ignored.

Property 1. The value of an edge between the same unique deployment in different planes is zero.

Due to the complexity of the N-dimensional QoS solution space in relation to component-based system deployment, we focus on evaluating QoS in a single dimension in this paper. We, therefore, narrow our focus to a 1-planar directed graph (Figure 4), which drastically reduces our solution space. The following section details how we can search this solution space to understand and locate deployments to improve systemic QoS properties along a single dimension.

3.2 Using Baseline Profiles to Search Deployment Solution Space

Each component in the component-based system produces a baseline profile that details its optimal performance. Such performance is realized when the component is deployed into a controlled and isolated environment, *i.e.*, no interference from other components in the system. As a component is collocated with other components, its performance will begin to deviate from its baseline performance. This is due to both software and hardware contentions.

Hardware contentions can be resolved using traditional analysis techniques, such as queuing theory [8], because they have more deterministic properties, such as service rate/time, that can be captured in a mathematical model. Software contentions, however, require execution in the target environment to fully locate and understand because it is hard to capture such problems using a model.

In Section 3.1, we discussed how the deployment solution space in a single QoS dimension could be represented as a directed graph G'. Each directed edge $e \in G'$ represents the difference in QoS performance between two unique deployments. When trying to locate candidate D&Cs that meet a specific QoS performance, or improve systemic QoS performance, it is ideal to locate deployments that yield performance metrics for each individual component that are close to their baseline. The problem becomes hard when all components and their interactions must be taken into account all at once.

We, therefore, use the percentage error (Equation 3) between observed and baseline performance of a component and deployment history to search the deployment solution space. We do not use the absolute value in the numerator when calculating the percentage error because the positive/negative sign signifies if QoS improved/worsened. Algorithm 1 lists our algorithm for locating a new deployment after labeling the current edge with the percentage difference in QoS between the previous deployment and the current deployment.

$$\% \text{ error} = \frac{\text{observed} - \text{baseline}}{\text{baseline}} \times 100 \tag{3}$$

If the current deployment has improved QoS performance in a single dimension, such as end-to-end response time, we use Algorithm 1 to locate a new deployment and continuing improving QoS. As listed in Algorithm 1, given the current graph of the D&C solution space, hosts and components in the system, and baseline and observed performance, we find the component with the max percentage error (Line 8). We then determine if we can relocate the component based on its history (Line 11), such as how often has it been relocated to a new host or how has it performed in previous tests on a given host with other components. If we can relocate the component i, we find a candidate host h to relocate the component to (Line 14).

hist and *candidate* in Algorithm 1 are placeholders for a domain-specific function that understands how to evaluate the history of the tests and locate a candidate host based on the previous and current test, respectively. Moreover, the *hist* and *candidate* functions are used to prevent a greedy approach to frequently selecting the same component and host, respectively, as targets for relocation—which we call *deployment thrashing*—and exhaustively searching the solution space[1].

If the new deployment is unique in G (Line 17) and does not create a cycle, then we add it to the graph and return the new deployment (Line 19). Lastly, we execute the new deployment and repeat the location process for the new deployment. In the case that

[1] It is possible to design a history and candidate function that exhaustively searches the solution space, however, such functions are not ideal as the solution space grows larger and more complex.

Algorithm 1. General algorithm for locating a new deployment in a single QoS dimension

```
1: procedure LOCATE(G, v, C, H, O, B)
2:     G: current graph of D&C solution space
3:     v: current location in graph G
4:     C: set of components in system
5:     H: set of hosts in system
6:     O: set of current observed performance for C
7:     B: set of baseline performance for C
8:     Δ = ∀c ∈ C : (o_c − b_c)/b_c × 100, o_c ∈ O, b_c ∈ B
9:     while Δ ≠ ∅ do
10:        Δ_i = max(Δ), i ∈ C
11:        if hist(G, i) < hist(G, (C − i)) then
12:            H' = H
13:            while H' ≠ ∅ do
14:                h = candidate(H', v, Δ)
15:                if H' − h ≠ H' then
16:                    e_{v,v'} = newedge(G, v, h, i)
17:                    if unique(G, v') then
18:                        accept(G, e_{v,v'})
19:                        return v'
20:                    end if
21:                    H' = H' − h
22:                else
23:                    H' = ∅
24:                end if
25:            end while
26:        end if
27:        Δ = Δ − Δ_i
28:    end while
29:    return NULL
30: end procedure
```

the current deployment does not improve the desired QoS, we backtrack to the previous deployment from the current deployment (vertex) and repeat the location process, *i.e*, Algorithm 1. We stop once we backtrack to the initial deployment and cannot locate a candidate component to reassign to a new host. This simple algorithm allows us—and in turn—developers to both understand the deployment solution space and locate solutions that meet QoS expectations.

4 Evaluating Our Deployment Search Technique Using the SLICE Case Study

In this section we present the results for searching the deployment solution space for the SLICE scenario using the search algorithm discussed in Section 3.

4.1 Experiment Design

The SLICE scenario introduced in Section 2.1 consists of 7 different component instances that must be deployed across 3 different hosts. Unlike any previous work that used the SLICE scenario [4], we do not place any constraints on how the components must be deployed onto their hosts. Instead, we are interested in locating deployments that yield better average end-to-end execution time for SLICE scenario's critical path—irrespective of the 350 msec deadline—based on Algorithm 1 discussed in Section 3.2.

The *hist* and *candidate* functions of Algorithm 1 are placeholders for domain-specific functions, *i.e*, functions that can better analyze the target domain and execution environment. Accordingly, we defined our history function for an individual component (Equation 4) and a set of components (Equation 5) as the percentage of times a component has been moved throughout the entire testing process.

$$hist(G,i) : \frac{\#\text{ times } i \text{ moved}}{\#\text{ of vertices in } G} \times 100 \quad (4)$$

$$hist(G,C) : \forall i \in C \frac{\#\text{ times } i \text{ moved}}{\#\text{ of vertices in } G} \times 100 \quad (5)$$

As shown later, this ensures that we do not constantly select the same component with the worst percentage error between unique deployments, and select components that may not have the worst percentage error in a test because it may not have been previously moved. The candidate function for selection is expressed in Equation 6,

$$candidate(H,v,\Delta) : min(avg(\Delta_{vh})) \quad (6)$$

where Δ_{vh} is the set of percentage error for components deployed on host $h \in H$ in deployment v, and identifies the host that contains the set of components with the least average percentage error from their baseline performance. This is conceptually similar to load balancing.

To test our search algorithm, we used the *Component Workload Emulator (CoWorkEr) Utilization Test Suite (CUTS)* [4], which is a system execution modeling tool for component-based system. CUTS enables developers to rapidly construct high-level behavior models of a system using domain-specific modeling languages [6] and generate source code for a target component middleware that can be executed and monitored in the target environment. For the target execution environment, we used ISISLab (`www.isislab.vanderbilt.edu`) at Vanderbilt University, which is a cluster of computers managed and powered by Emulab [16] software. Table 2 lists the specifications of the three hosts we used in ISISLab for our experiments.

Table 2. Host specification for experiments

ISISLab Host	Specification
1, 2, 3	Fedora Core 4, dual core 2.8 GHz Xeon 1 GB RAM, 40 GB HDD, 4 Gbps NIC

Table 3. Measured baseline of SLICE components

Component	Input	Output	Baseline (ms)
(A) Config	Assess	Command	10
(B) EffMain (C) EffSec	Command	Status	10
(D) PlanOne	Track	Situation	45
(E) PlanTwo	Situation	Assess	45
(F) SenMain (G) SenSec	Command	Track	45

We used the three ISISLab hosts to measure the baseline performance of each component on each host in an isolated and controlled environment. This is necessary because each component's baseline performance is needed for Algorithm 1 in Section 3.2. Each baseline test was executed for 2 minutes, which resulted in 120 generated events handled by each component, because 2 minutes was enough time for the component's baseline performance to stabilize.[2] Table 3 lists the baseline metric for each component by their input port (*e.g.*, event sink) and output port (*e.g.*, event source). We consolidated the baseline performances for each component on each host because they were equal.

4.2 Experiment Results

Interpreting the performance results. There are $3^7 = 2187$ unique deployments in the SLICE scenario. Table 4 lists the results for testing 17 unique deployments out of the 2187—for the SLICE scenario. The 17 unique deployments were derived using the search algorithm presented in Section 3.2 and Equations 4, 5, and 6. Out of the 17 unique deployments, we located 6 deployments (*i.e.*, tests 2, 4, 7, 15, 16, and 17) that improved the critical paths end-to-end execution time from the initial deployment.

The number of unique deployments in the SLICE scenario has never been exhaustively tested. We, however, can estimate how good a deployment is by summing the baseline performance for all the components in an execution path— assuming network and queuing performance is negligible since they represent another dimension of QoS—and comparing it against its observed performance. This summation represents the optimal performance for an execution path given there is no network contention and no queuing overhead.

The sum of the component's performance in the critical path for the SLICE scenario (see Section 2.1) is 155 msec. This is derived by summing the baseline performances for all the components in the critical path, which is given in Table 3. Test 4, 7, 15, 16, and 17 were the closest test to the optimal baseline end-to-end response time, *i.e.*, $< 16\%$ deviation, with test 17 being the closest, *i.e.*, a 12% deviation.

We also observed that for all 17 tests, the best end-to-end response time performance for the critical path was 169 msec (not present in Table 4). Even when all components

[2] The duration of a baseline test depends on the complexity of the component's behavior, such as (non-)linear and conditional workflows.

Table 4. Results for searching SLICE deployment solution space

Test	Deployment Strategy			Avg. End-To-End
	Host 1	Host 2	Host 3	Performance (ms)
1	B,D,F,G	A	C,E	182.7
2	A,B,D,F,G		C,E	181.2
3	A,B,F,G,D	E	C	185.9
4	B,C,D,F,G	C	E	180.2
5	A,D,F,G	C	B,E	186.7
6	A,B,D,G	C	E,F	186.5
7	A,B,D,F	C	E,G	180.4
8	A,B,F,G	C	D,E	193.3
9	A,B,C,D,F,G		E	182.5
10	B,D,F,G	C	A,E	189.6
11	A,D,F,G	B,C	E	206.9
12	A,B,D,G	F	C,E	207.9
13	A,D,F,G	B	C,E	189.8
14	A,B,F,G	D	C,E	188.8
15	A,D,B,F	G	C,E	178.9
16	A,B,D,F		C,E,G	179.1
17	A,B,D,E,F	G	C	174.6

were deployed onto the same hosts, such as in test 17, the end-to-end response time was greater than the optimal. This is, therefore, evidence of some form of software contention because the event generation rate for the SLICE scenario, *i.e*, 1 second, is not high enough to produce a workload that causes over-utilization of resources, such as the CPU.

Understanding host selection. Table 4 shows that a majority of the unique deployments did not utilize Host 2. This is attributed to Host 2 always having the worst average percentage error for each test. We did observe similar performance results in our baseline test, however, our technique tries to create deployments where components perform optimally (*i.e.*, close to their baseline) when collocated with other components. Moreover, our candidate function (Equation 6) was defined the avoid hosts that had components with a high percentage error from their baseline performance. We, therefore, did not investigate this anomaly and interpreted such results as recognizing and avoiding "bad" hosts that could potential worsen QoS performance.

Interpreting the search graph. Figure 6 is the partial graph, or ordered tree, for the tests presented in Table 4. It illustrates the search process of Algorithm 1 using Equations 4, 5, and 6 for their corresponding placeholders in Algorithm 1. As illustrated in Figure 6, we first started with a random deployment of the SLICE scenario. The second test improves end-to-end response time for the critical path, however, the third test does not improve end-to-end response time. We, therefore, backtracked to the second test and selected a new component to relocate to a different host. This resulted in deriving test 4, which improved end-to-end response time for the critical path. Test 7 had an end-to-end

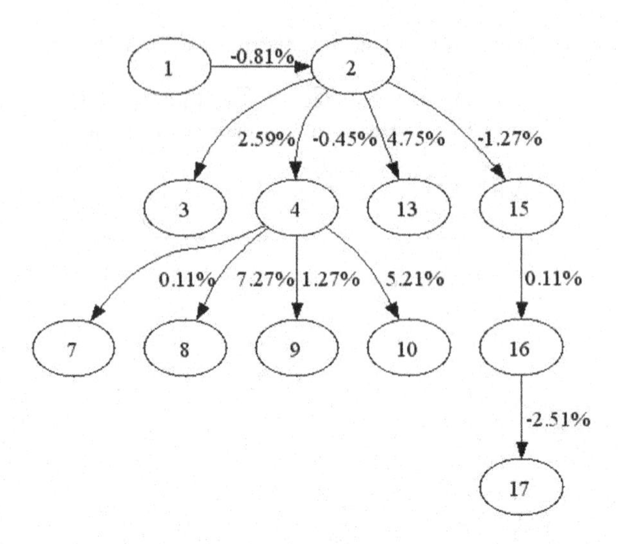

Fig. 6. Partial search graph for SLICE's unique deployments

response time close to test 4 (*i.e.*, 0.11% difference), but it was not accepted since it did not improve end-to-end response time. It is, however, possible to use a standard error to determine the probability of accepting QoS performances that are close in value.

After test 7, we did not locate a new deployment that improved the critical path's end-to-end response time until test 15. This was a result of backtracking to the unique deployment for test 2, and selecting a new component to relocate to a new host. Consequently, we derived unique deployments for test 15, an improvement to test 2; test 16, an improvement to test 15; and test 17, an improvement to test 16.

We, therefore, can conclude that our algorithm can be parameterized with domain-specific functions to logically search the deployment solution space for a component-based system, and locate deployments that will improve QoS in a single dimension, such as average end-to-end response time, based on component baseline profiles. More importantly, it is possible to locate deployments that improve QoS in a single dimension based on a component's baseline profile.

5 Related Work

Traditional analytical techniques, such as rate monotonic analysis [14], queuing networks [15], and queuing petri nets [5], have been used to predict the performance of DRE systems. Such techniques enable system developers to predict performance at design time, however, they require in depth knowledge of the system architecture. Moreover, these techniques do not completely take into account the target operating environment, unless such a model exists and has been verified. Our technique differs from traditional analytical techniques because we evaluate a deployment by executing it in the target environment. We, therefore, are able to take into account the entire operating environment irrespective of system developers having any knowledge of its properties.

We also believe that such analytical techniques can be integrated with our approach to analyze candidate deployments before executing them in the target environment to reduce search complexity and false positives.

Diaconescu et al. [2] discusses a framework and technique for improving the performance of component-based system by selecting implementations that are more suitable for a given execution context. Their technique involves "training" an adaptation module, locating performance anomalies in deployed applications, and select implementations that will address the anomaly and improve performance. Our technique is similar in that we use baseline profiles, which can be viewed as "training" to understand component performance. Our approach differs in that we do not focus on run-time adaption to improve performance. Instead, we are focus on locating a collection of unique deployments at design-time that will improve QoS performance, such as end-to-end response time.

Memon et al. [7] discuss a framework and customizable technique for providing continuous quality assurance for applications with a large configuration space. Their technique randomly tests different solutions in the configuration space and when a valid configuration, *i.e.*, one that meets their goal, is found, other configurations related to the good configuration is tested. Our approach is similar to their approach in that our search functions are customizable. Our approach differs in that we do not permutate all possible unique deployments and randomly test until we find a valid deployment. Instead, we start with a random deployment and base subsequent searches on the observations of the current test. Moreover, we base our search criteria on a component's percentage error from its baseline performance instead of testing by solely moving a component to a new host.

6 Concluding Remarks

Bad deployment choices for a component-based system can have dire effects on systemic QoS properties, such as end-to-end response time. In this paper, we presented a simple technique for locating deployments to improve system QoS along a single dimension of QoS. Our technique uses baseline profiles to calculate the percentage error in a component's performance irrespective of resource utilizations. This performance metric is used in conjunction with domain-specific functions to evaluate the history of a component and locate candidate hosts for relocating individual components based on a component's percentage error and current state of testing.

This approach allows for a flexible technique for searching the deployment solution space for systems that do not suffer from high resource utilization problems. Likewise, as more is learned about the system (and domain), system developers can modify their domain-specific functions to improve search capabilities.

Lessons Learned and Future Research

Techniques for searching deployment solution space is a process that continuously evolves as more is learned about the domain. The following list, therefore, discuss the lessons learned and future research directions:

- The SLICE scenario has 2187 unique deployments, however, the solution space has not been exhaustively tested. Future work, therefore, includes exhaustively testing the SLICE scenario's deployment solution space to locate the optimal deployments under various workloads, such as increased event generation rate, to strengthen the validation of our technique.

- Manually searching the deployment solution space, regardless of any search algorithm used, is inefficient and error prone. Future work, therefore, includes extending CUTS to automate the process of intelligently searching the solution space, which we believe will help address scalability challenges (*i.e.*, searching solution spaces where $|C|$ and $|H|$ are large numbers); however addressing scalability is highly dependent on the domain-specific functions ability to identify valid components and hosts for relocation.
- In some situations, exhaustively searching the deployment solution space many be less expensive than searching the solution space using Algorithm 1 in Section 3.2. This situation is highly dependent on the system under evaluation. Future work, therefore, including identifying and understanding situations where exhaustively searching the solution space is more favorable.
- Currently, we are testing unique deployments in an isolated environment. In a realistic environment, however, an application may be deployed with other applications. Future work, therefore, includes extending our technique to include situations of improving QoS in a single dimension when the DRE system comprises components belonging to multiple applications that are executing in the same environment.
- The generalized QoS problem has a N-dimensional solution space, however, we reduced the solution space to one dimension in this paper to simplify the deployment search problem and focus on the technique. Future work, therefore, will extend our technique to include multiple dimensions of QoS.

The CUTS framework and our research artifacts are available in open source at `www.dre.vanderbilt.edu/CUTS`.

References

1. Deng, G., Balasubramanian, J., Otte, W., Schmidt, D.C., Gokhale, A.: DAnCE: A QoS-enabled Component Deployment and Configuration Engine. In: Dearle, A., Eisenbach, S. (eds.) CD 2005. LNCS, vol. 3798, pp. 67–82. Springer, Heidelberg (2005)
2. Diaconescu, A., Murphy, J.: Automating the Performance Management of Component-based Enterprise Systems Through the Use of Redundancy. In: Proceedings of the 20th IEEE/ACM International Conference on Automated Software Engineering (ASE 2005), pp. 44–53 (2005)
3. Hill, J.H., Gokhale, A.: Model-driven Specification of Component-based Distributed Real-time and Embedded Systems for Verification of Systemic QoS Properties. In: Proceeding of the Workshop on Parallel, Distributed, and Real-Time Systems (WPDRTS 2008), Miami, FL (2008)
4. Hill, J.H., Slaby, J., Baker, S., Schmidt, D.C.: Applying System Execution Modeling Tools to Evaluate Enterprise Distributed Real-time and Embedded System QoS. In: Proceedings of the 12th International Conference on Embedded and Real-Time Computing Systems and Applications, Sydney, Australia (2006)

5. Kounev, S.: Performance Modeling and Evaluation of Distributed Component-Based Systems Using Queueing Petri Nets. IEEE Transactions on Software Engineering 32(7), 486–502 (2006)
6. Ledeczi, A., Bakay, A., Maroti, M., Volgysei, P., Nordstrom, G., Sprinkle, J., Karsai, G.: Composing Domain-Specific Design Environments. IEEE Computer, 44–51 (2001)
7. Memon, A., Porter, A., Yilmaz, C., Nagarajan, A., Schmidt, D.C., Natarajan, B.: Skoll: Distributed Continuous Quality Assurance. In: Proceedings of the 26th IEEE/ACM International Conference on Software Engineering, Edinburgh, Scotland (2004)
8. Menasce, D.A., Dowdy, L.W., Almeida, V.A.F.: Performance by Design: Computer Capacity Planning By Example. Prentice Hall PTR, Upper Saddle River (2004)
9. de Niz, D., Rajkumar, R.: Partitioning Bin-Packing Algorithms for Distributed Real-time Systems. International Journal of Embedded Systems (2005)
10. Parsons, T., Murphy, J.: Detecting Performance Antipatterns in Component Based Enterprise Systems. Ph.D. thesis, University College Dublin, Belfield, Dublin 4, Ireland (2007)
11. Rajkumar, R., Lee, C., Lehoczky, J., Siewiorek, D.: A Resource Allocation Model for QoS Management. In: Proceedings of the IEEE Real-Time Systems Symposium (1997)
12. Shankaran, N., Koutsoukos, X., Schmidt, D.C., Gokhale, A.: Evaluating Adaptive Resource Management for Distributed Real-time Embedded Systems. In: Proceedings of the 4th Workshop on Adaptive and Reflective Middleware, Grenoble, France (2005)
13. Smith, C., Williams, L.: Performance Solutions: A Practical Guide to Creating Responsive, Scalable Software. Addison-Wesley Professional, Boston (2001)
14. Tri-Pacific: RapidRMA (2002), http://www.tripac.com
15. Ufimtsev, A., Murphy, L.: Performance Modeling of a JavaEE Component Application using Layered Queuing Networks: Revised Approach and a Case Study. In: Proceedings of the Conference on Specification and Verification of Component-based Systems (SAVCBS 2006), pp. 11–18 (2006)
16. White, B., Lepreau, J., Stoller, L., Ricci, R., Guruprasad, S., Newbold, M., Hibler, M., Barb, C., Joglekar, A.: An integrated experimental environment for distributed systems and networks. In: Proc. of the Fifth Symposium on Operating Systems Design and Implementation, pp. 255–270. USENIX Association, Boston (2002)

Preserving Intentions in SOA Business Process Development

Lucia Kapová[1,3], Tomáš Bureš[1,2], and Petr Hnětynka[1]

[1] Department of Software Engineering, Faculty of Mathematics and Physics, Charles University
Malostranske namesti 25, Prague 1, 118 00, Czech Republic
{kapova,bures,hnetynka}@dsrg.mff.cuni.cz

[2] Institute of Computer Science, Academy of Sciences of the Czech Republic
Pod Vodarenskou vezi 2, Prague 8, 182 07, Czech Republic

[3] Software Engineering, FZI Forschungszentrum Informatik
76131 Karlsruhe, Germany

Summary. Business processes play an important role in Service-Oriented Architectures. Commonly, the business processes are designed in the Business Process Modeling Notation (BPMN), which allows their development even by persons without programming skills. Being abstract and high-level BPMN is not suitable for direct execution, though. The natural choice for implementing business processes is the Business Process Executable Language (BPEL), which is directly executable, but it is also a programming language that requires programming skills in order to be used. This dichotomy is typically solved by transforming BPMN into BPEL. However, this transformation is a complex task. There have been developed a number of transformation methods but their common problem is either incompleteness or loss of intentions, which makes BPEL rather difficult to modify and debug as well as to propagate changes back to BPMN. In this paper we address this problem by presenting a non-trivial enhancement of the existing transformation algorithm [14]. Our approach provides a complete transformation while preserving a large set of intentions expressed in the BPMN description, which makes it also suitable in the context of model-driven development.

1 Introduction

Service Oriented Architectures (SOA) have become an established methodology in the enterprise IT world. SOA speeds up and simplifies development of large enterprise systems by allowing for composing the target system from composable services with well-defined interfaces. These services may be implemented in various technologies and programming languages (e.g. Java, .NET). Another very important concept of SOA is "orchestration". In simplicity, this term denotes a piece of prescription that describes how to perform a complex task with the help of other services. Orchestration is typically described by a dedicated language, most predominantly BPEL [10]. In a sense, a BPEL script specifies a workflow which tells the order of other services' invocation and how to aggregate the results. Of course, BPEL, being a full-fledged programming language, may additionally contain control-flow statements, simple computations, etc.

R. Lee (Ed.): Soft. Eng. Research, Management & Applications, SCI 150, pp. 59–72, 2008.
springerlink.com

Another SOA's primary objective is automation of business processes. Although SOA is sometimes confused with business processes, it is generally not possible to put the equals sign between them. Business processes comprise much wider area, in addition to software systems they encompass interaction with and among humans, organizational units, etc. The proper relation is that SOA may be used to automate a part of a business process.

Next important distinction between business processes and SOA (BPEL in particular) is that business processes are typically created by persons with managerial skills and deep knowledge in their domain but with only user-level knowledge of using computers. The most formal description that is obtainable here is typically a graphical capture of the processes by a workflow in BPMN [19].

On the other hand, implementing SOA requires programming skills. For example, BPEL, which is one of the most high-level and abstract descriptions used in SOA, is an XML-based programming language requiring the knowledge of a lot of low-level details (such as transport protocols, XML namespaces, XSchema types, etc.). The work with BPEL may be facilitated by using existing tools and frameworks such as IBM Web-Sphere Choreographer [6] and Oracle BPEL Process Manager [13]. However, these tools do not go further than providing graphical editing of the BPEL structure and simplifying deployment of a BPEL script.

The bottom line is that there are two worlds which co-exist: (a) persons with managerial skill and expertise in their domain but with little knowledge of programming, and (b) IT-specialists, who are able to implement a part of a business process in SOA (BPEL in particular) but who do not have the proper knowledge of the business processes nor of their management.

It is obvious that such a dichotomy may cause problems when trying to automate business processes using SOA. A remedy in this case is provided by the approach of Model Driven Architectures (MDA [11]). MDA allows gradually refining a particular model going from a Platform Independent Model (PIM) to a Platform Specific Model (PSM). In the context of business processes and SOA, an abstract BPMN workflow may be regarded as PIM and an executable BPEL script as PSM. The MDA's approach is then to transform BPMN to BPEL.

The transformation of BPMN to BPEL needs a lot of added low-level information (which is not present in the original BPMN) and thus the transformation is more-or-less manual. However, there are attempts, which automate it to some extent (see Section 1.2). These attempts rely on mapping BPMN constructs on BPEL ones. The BPEL code is then further refined in existing BPEL tools. Bindings to other services are added, the BPEL script is debugged and then eventually deployed.

1.1 Goals and Structure of the Paper

The problem of transforming from BPMN to BPEL is rather a challenging issue. BPMN is an unstructured graph oriented language. BPEL, on the other hand, is a well-structured language with a lot of low-level details. The key problem is that the transformation should preserve the intentions, which the person describing the process has expressed using the BPMN workflow. It is necessary to recognize particular patterns in BPMN and map them to corresponding BPEL constructs.

The goal of this paper is an extension of the general ability of transforming BPMN-to-BPEL while preserving the intentions and producing easily understandable output. We focus on a frequent pattern of *partially synchronized workflow threads.* In this pattern there are a number of concurrent threads which have a few uni-directional control-flow dependencies – meaning that an action of one thread may commence only after another action in another thread has been completed (this is exemplified later in the text in Figure 2).

We base our approach on an existing algorithm [14], which performs the BPMN-to-BPEL translation in a generic way. We enrich it to explicitly support the pattern mentioned above.

The paper is structured as follows. The rest of this section presents other approaches related to our solution. Section 2 describes the original transformation algorithm and analyzes its issues on an example. In Section 3, we introduce the proposed extension of the original algorithm and apply resulting algorithm on the case study. Finally, Section 4 concludes the paper and provides future directions.

1.2 Related Work

The transformation of a workflow to BPEL has been to some degree in focus of several works. The transformation is typically connected with identification and classification of workflow patterns, which enables the tranformation to preserve original intentions and map the workflow patterns to appropriate BPEL constructs. In this respects works [17, 8] provide a very good basis. The paper [17] concentrates mainly on identification and description of various types of workflows, while [8] deals with unstructured workflows and provides charaterization of correct and faulty workflows.

The transformation to BPEL itself has been studied on special kind of Petri Nets, so called WorkflowNets [1]. A tool transforming the WorkflowNets to BPEL has been developed too [7].

Another formalism for capturing workflows – UML Activity Digrams [12] – has been used for example in [4]. In this work, the authors present a way of deriving BPEL from the Activity Diagrams. The same formalism is used in [5], which focusses specifically on untangling loops in the context of BPEL.

Regarding the BPMN-to-BPEL transformation in particular, this has been pioneered by informal case-studies such as [18], which provides an example of a transformation of a complex BPMN workflow to a BPEL process. The transformation in general has been discussed in works [16, 9], which point out conceptual mismatches between BPMN and BPEL and discuss the main general transformation strategies.

In our approach we take up as a basis an existing algorithm described in [14] and implemented in a tool called BABEL (see [14] too). The algorithm has been further extended in [15], where the authors extend the set of covered patterns, however still not covering the pattern of *partially synchronized workflow threads* which we deal in this paper with.

In fact, the pattern of *partially synchronized workflow threads* is not supported by any of the approaches transforming a workflow to BPEL. The existing approaches convert this pattern usually to a set of BPEL event handlers. In the worst case, every edge in the workflow is encoded as an event. That yields a valid and functional BPEL, but the original intentions are lost completely. The resulting BPEL script is thus incomprehensible and tasks as refining it or debugging it are out of question. Also a backward transformation (from BPEL-to-BPMN), which is necessary to keep the original design in sync with the actual implementation (so called *round-trip engineering* [3]), is made very complicated by this loss of the intentions. In fact the backward transformation is one of the essential features needed for proper support of MDA.

2 The Original Transformation Algorithm

We present a brief overview of the original transformation algorithm in this section. The input of the algorithm is a Business Process Diagram (*BPD*) defined using the *core* subset of BPMN (see Figure 1). There are *objects*, like *event* (signal), *task* (atomic activities) or *gateway* (routing construct) considered, and *sequence flow*, representing the control flow relation.

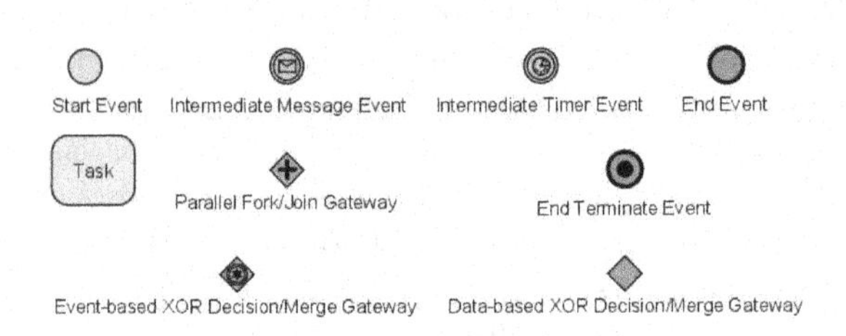

Fig. 1. The core BPMN

The *BPD* defines a *directed graph* with *vertices* (objects) O and *edges* (sequence flows) F. In simplified notation it is represented as $BPD = (O, F, Cond)$, where $Cond$ is a boolean function representing condition associated with a *sequence flow*. An input of the transformation algorithm has to be *well-formed BPD*, which is *BPD* restricted by these requirements: (a) start events have an indegree of zero and an outdegree of one; (b) end events have an outdegree of zero and an indegree of one; (c) sets of start and end events are considered as singletons; (d) tasks and intermediate events have an indegree of one and an outdegree of one; (e) fork or decision gateways have an indegree of one and an outdegree of more than one; (f) join or merge gateways have an outdegree of one and an indegree of more than one. (For simplicity, we have considered only related parts of the definition, for details see [14].)

Transformation of the *directed graph* defined by *BPD* to the block-structured target language is done through decomposing *BPD* into *components*. The term *component* is used to refer to a subset of *BPD*.

Definition 1 (Component). *$C = (O_c, F_c, Cond_c)$ is a component of a well-formed $BPD = (O, F, Cond)$ if all of the following hold: (a) there is a single entry point outside the components; (b) there is a single exit point outside the components; (c) there exist a unique source object and a unique sink object; (d) $F_c = F \cap (O_c \times O_c)$; (e) Cond function domain is restricted to F_c.*

The basic idea of the transformation algorithm is to select *components* in the *BPD*, provide its BPEL translation and *fold* the *components*. *Components* are distinguished between *well-structured* and *non-well-structured*. First type of *components* can be directly mapped onto BPEL structured activities (*sequence*, *flow*, *switch*, *pick* and *while*), the second one can be translated into BPEL via *event-action rules*, for details see [14]. Identification of *components* starts with maximal *sequence* identification, other structured *components* identification and at last minimal *non-well-structured components* are identified. In this way identified *component* is then replaced by function *Fold* to a single task object in *BPD*. Function *Fold* is called until there is no other *component* identified. *BPD* without any *component* is called a *trivial BPD*. Translating of *trivial BPD* is straightforward using BPEL `process` constructs.

The so-called event-action rule-based transformation approach to *non-well-structured component* can be applied to translating any *component* to BPEL, however with the drawback of loosing intentions. As the first step of this transformation, a set of preconditions for each object within a *component* is defined. Each set of preconditions captures all possible ways that lead to the execution of an associated object. In the second step precondition sets are transformed with their associated objects into a set of event-action rules representing on which event is the rule triggered and what actions are performed. The last step is to translate event-action rules to BPEL code. Each set of event-action rules that was derived from a *component* is translated to a BPEL scope that defines execution of certain actions on events by event-handler association.

The original algorithm to translate a well-formed core BPD into BPEL is defined as follows (the set of *components* of a *BPD* (X) is denoted as $[X]_c$).

Algorithm 1 (Original Transformation Algorithm). *Let BPD be a well-formed core BPD with one start event and one end event.*

1. *$X := BPD$*
2. *if $[X]_c = \emptyset$ (i.e., X is a trivial BPD), output the BPEL translation of the single task or event object between the start and end events in X, goto 5.*
3. *while $[X]_c \neq \emptyset$ (i.e., X is a non-trivial BPD)*
   a) *if there is a maximal SEQUENCE-component $C \in [X]_c$, select it and goto 3d.*
   b) *if there is a well-structured component $C \in [X]_c$, select it and goto 3d.*
   c) *if there is a minimal non-well-structured component $C \in [X]_c$, select it and goto 3d.*

d) Attach BPEL translation of C to task object t_c.
e) $X := Fold(X, C, t_c)$ and return to 3.

4. *Output the BPEL code attached to the task object t_c.*
5. *A start event and a end event are translated into a pair of* `<process>` *and* `</process>` *tags to enclose the BPEL code generated in steps 2 or 5. For a terminate event,* `<exit/>` *activity is added.*

2.1 Transformation Example

As a running example, we use the initial example from the BPEL specification [10], which is a Purchase Order process (see Figure 2). The process is initiated by receiving the purchase order message from a customer. Next, three concurrent *sequences* are initiated: calculating the price of the order (Tasks T3 and T6), selecting a shipping company (Tasks T2 and T4), and scheduling the production (Tasks T1 and T5) and the shipment for the order (Task T7). There are control and data dependences between these *sequencies* (e.g., the shipping price or shipping date is required to complete the price calculation or scheduling task). Eventually an invoice is processed and sent to the customer.

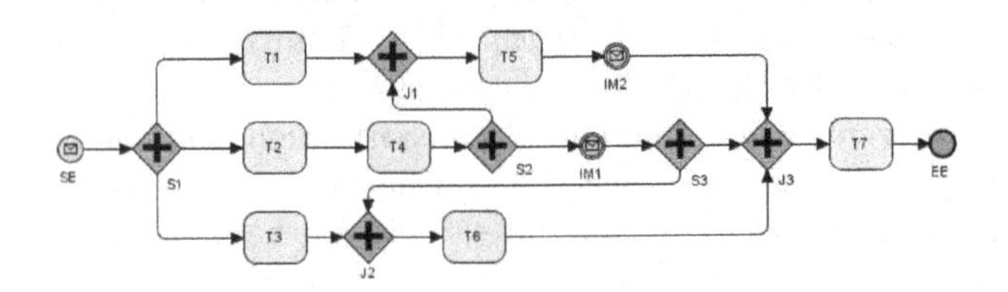

Fig. 2. Business process model

When processing this example with the algorithm, it first identifies *components* in the model. It finds maximal sequences C_1 and C_2, followed by *non-well-structured component* C_3 and eventually by the maximal sequence C_4 (see Figure 3).

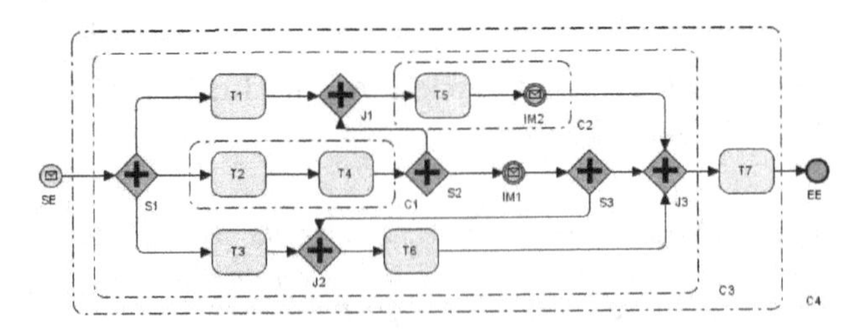

Fig. 3. Transformation steps – components identification

After identifying each of the components, function *Fold* is called. This results gradually in BPDs as illustrated in Figure 4 and Figure 5.

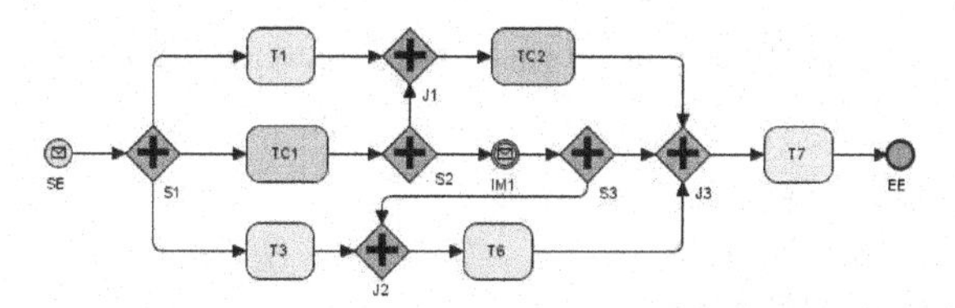

Fig. 4. Transformation steps – after folding components C_1 and C_2

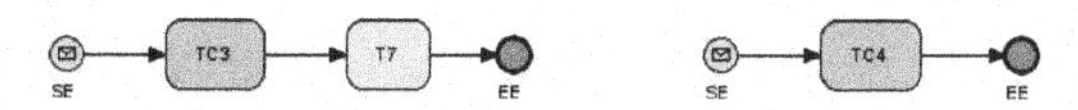

Fig. 5. Transformation steps – after folding component C_3 and the final result

Finally, is a BPEL translation attached to each identified *component.* The *components* C_1 and C_2 are translated to:

```
<sequence name="TC1">
  <invoke name="T2" .../>
  <invoke name="T4" .../>
<sequence/>

<sequence name="TC2">
  <invoke name="T5" .../>
  <recieve name="IM2" .../>
<sequence/>
```

For translation of *component* C_3, the application of event-action rule-based transformation approach is needed. The resulting code is generated within a scope construct. Short example follows.

```
<scope name="TC3">
  <onEvent Start(C3)>
    <flow name="F1">
      <invoke flow(S1,T1)/>
      <invoke flow(S1,C1)/>
      <invoke flow(S2,T3)/>
    </flow>
  </onEvent>
  <onEvent flow(S1, T1)>
    <invoke end(T1)/>
  </onEvent>
  <onEvent end(T1)>
    ...
  </onEvent>
  ...
  ...
  <invoke Start(C3)/>
</scope>
```

The whole process is then generated as the sequence TC_4 enclosed in BPEL `<process>` tags:

```
<process name="purchase order">
  <sequence name="TC4">
    <scope name="TC3"/>
    <invoke name="T7" .../>
  </sequence>
</process>
```

The problem is that the algorithm translates the dependencies among sequences to inappropriate `onEvent` constructs. An correct and intention preserving transformation would be to identify the synchronization links in the workflow and map them to BPEL `link` constructs, which serve exactly for this task.

3 Algorithm Extension

The pattern of *partially synchronized workflow threads*, which we have presented in Section 2.1 (BPD component C_3), is quite typical. In this paper we thus aim at extending the original algorithm [14] by the support for this pattern. We identify the synchronization links and map them to the corresponding BPEL `link` constructs. Having done this, we are able to get more readable, comprehensible and intentions preserving results.

The algorithm extension operates on a *non-well-structured component* which starts with a fork gateway and ends with a join gateway (e.g. component C_3 in Figure 2), and tries to identify *execution paths* in the flow. In other words it tries to map the component to the following BPEL construct.

```
<flow>
  <links>
    ...
  </links>

  <sequence>
    ...
  </sequence>
  <sequence>
    ...
  </sequence>
</flow>
```

Each execution path is reflected by the respective `sequence`. The synchronization among the sequences is performed by respective `links`, which are declared in the upper part of the previously shown `flow` construct. Figure 6 shows this on

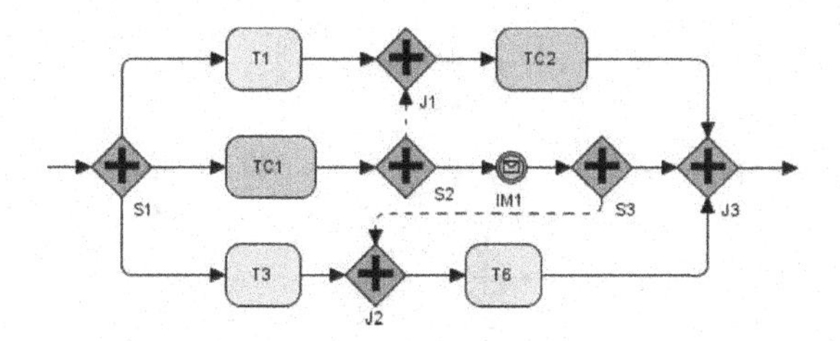

Fig. 6. Execution paths and synchronization links

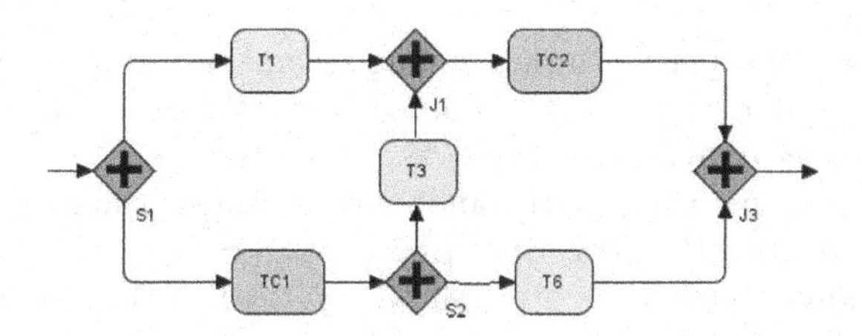

Fig. 7. Non-well-structured component not following the required pattern

the running example; solid lines denote the exection paths, dashed lines stand for the synchronization.

In some cases it is not possible to transform the *non-well-structured component* to flow with sequences which synchronize among one another. An example of a component where this transformation is not possible is in Figure 7. In such cases we resort to the more general way of mapping a *non-well-structured component* to BPEL which relies on the `onEvent` constructs as in the original algorithm.

3.1 Identifying Execution Paths

To identify the execution paths, it is necessary to decide in each fork gateway which branch is the execution path. The other branches would be synchronization links (see Figure 6).

Obviously the synchronization links may lead only between fork and join gateways. If there is a link starting or ending in a task, intermediate message event, initial fork or terminal join (e.g. S1 and J3 in Figure 6) it has to be part of the execution path.

However, there may be a number of fork and join gateways mutually interconnected serving thus for complex synchronization among several execution paths. To cope with this complexity, we introduce a concept of a *Hypergateway*, which we define as follows.

Definition 2 (Hypergateway). *A hypergateway is a maximal connected subgraph of a component which consists only of fork and join gateways.*

The hypergateway encapsulates neighboring fork and join gateways thus forming a gateway with several incoming links and several outgoing links (see Figure 8). It

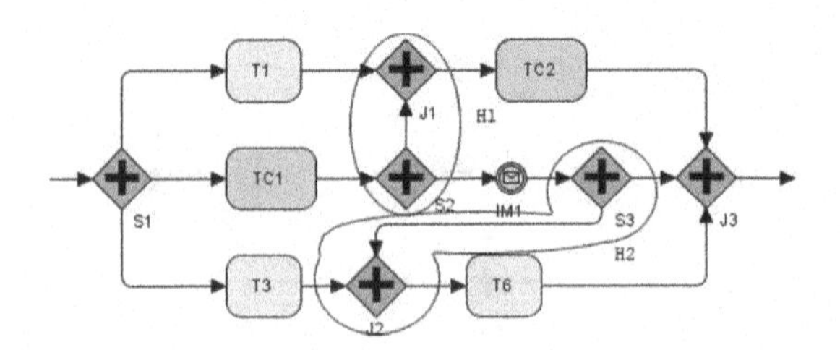

Fig. 8. Identification of hypergateways

gives a big picture of synchronization among several tasks, intermediate events, etc. Therefore the first step of the mapping a *non-well-structured component* to BPEL is identifying its hypergateways.

In the next step, execution paths and synchronization links inside each hypergateway are identified. This is done as follows.

A hypergateway is viewed as a graph $G(V, E)$, where V is a set of vertices that comprises fork and join gateways in the hypergateway and E as a set of edges contains links among the join and fork gateways.

For each incoming link i (from a task, intermediate message event, etc.) to the hypergateway, a vertex v_i^I is added to V and an edge (v_i^I, v^G) is added to E, where v^G corresponds to the fork/join gateway in the hypergateway which the incoming link leads to.

The same is done for each outgoing link j (to a task, intermediate message event, etc.) from the hypergateway. A vertex v_j^O is added to V and an edge (v^G, v_j^O) is added to E, where v^G corresponds to the fork/join gateway in the hypergateway which the outgoing link leads from.

Finally, vertices v^S and v^E are added to V and edges (v^S, v_i^I) and (v_j^O, v^E) are added to E for each incoming link i to the hypergateway and for each outgoing link j from the hypergateway. Figure 9 shows the resulting graph for the hypegrateway H_1 from the running example (Figure 8).

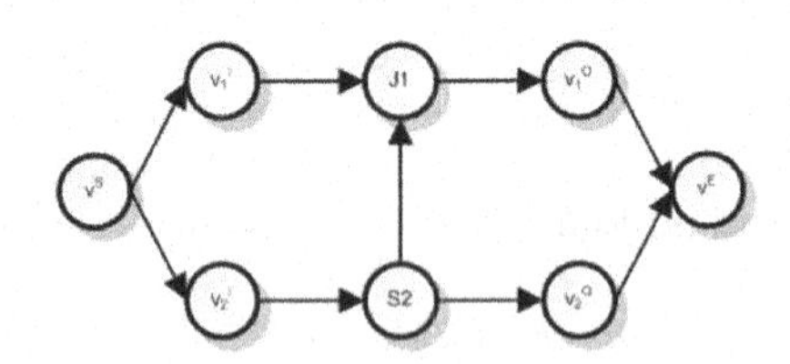

Fig. 9. Graph representation of hypergateway H_1

Identification of execution paths and synchronization links is now performed by finding the maximal flow in this graph. Every edge is assigned the capacity 1. Vertex v^S is the source and vertex v^E serves as the sink. Edges with actual flow of 1 constitute the execution paths. Edges with actual flow of 0 serve for synchronization (see Figure 10).

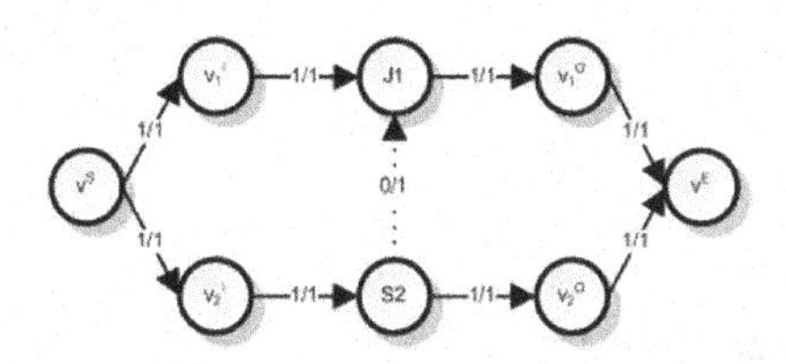

Fig. 10. Maximal flow computed for hypergateway H_1

For this procedure to work, it is needed that the maximal flow found assigns flow either 0 or 1 (i.e., not fractions). Fortunately, this is an inherent feature of the most commonly used algorithms for finding the maximal flow, such as Ford-Fulkerson or Edmonds-Karp [2].

Edges with the flow of 1 constitute disjunctive paths (not sharing edges or vertices); each one connecting an incoming link to the hypergateway with an outgoing link from the hypergateway. The disjunctiveness is assured by the fact that the fork and join gateways inside the hypergateway must have exactly either one incomming edge (for forks) or one outgoing edge (for joins).

By performing this procedure to each hypergateway, execution paths and synchronization links in the whole *non-well-structured component* are found. Cases when a *non-well-structured component* cannot be transformed to a flow with synchronized sequences are discovered during the computation of maximal flows in hypergateways. If the number of incoming and outgoing links of a hypergateway is not equal or if the value of the maximal flow is less than the number of links then the component does not correspond to the expected pattern and has to be represented using `onEvent` constructs.

3.2 Transformation to BPEL

Having the knowledge of execution paths and synchronization links, the transformation to BPEL is rather straightforward.

The *non-well-structured component* is (as exemplified previously) transformed to the BPEL `flow` construct. The execution paths are transformed to the BPEL `sequence` constructs and synchronization links are turned to the BPEL `link`, `sources`, and `targets` constructs.

For every synchronization link, a name is generated and the declaration of synchronization links is put into the top-level BPEL `flow` construct. The actual synchronization is performed by including `sources` and `targets` constructs into task invocations (performed by `invoke`) and intermediate message reception (performed by `receive`).

The `sources` construct lists links for which completing a task or reception of a message fires an event. The `targets` construct lists links with which execution of a task is synchronized.

The use of these constructs is well demonstrated in the following BPEL code, which outlines the result of transforming the *non-well-structure component* C_3 using the proposed algorithm extension.

```
<process>
 ...
  <sequence>
    <flow>
      <links>
        <link name="S2J1" .../>
        <link name="S3J2" .../>
      </links>
      <sequence>
        <invoke name="TC1" ...>
          <sources>
            <source linkName="S2J1" .../>
          </sources>
        </invoke>
        <receive name="IM1" ...>
          <sources>
            <source linkName="S3J2" .../>
          </sources>
        </receive>
      </sequence>
      <sequence>
        <invoke name="T1" ...>
          <targets>
            <target linkName="S2J1" .../>
          </targets>
        </invoke>
        <invoke name="TC2" .../>
      </sequence>
      <sequence>
        <invoke name="T3" ...>
          <targets>
            <target linkName="S3J2" .../>
          </targets>
        </invoke>
      </sequence>
    </flow>
    <invoke name="T7" .../>
  </sequence>
 ...
</process>
```

3.3 Integration with the Original Algorithm

Integrating the proposed extension to the original algorithm [14] is relatively simple. We extend the step 3c (handling of a *non-well-structured component*) by performing the proposed transformation first. If the transformation discovers that the component does not correspond to the required pattern (flow with synchronizing sequences), it falls back to the original solution for *non-well-structured components.* The overall algorithm thus looks the following way.

Algorithm 2 (Extended Transformation Algorithm)
Let BPD be a well-formed core BPD with one start event and one end event.

1. $X := BPD$
2. *if* $[X]_c = \emptyset$ *(i.e., X is a trivial BPD), output the BPEL translation of the single task or event object between the start and end events in X, goto 5.*
3. *while* $[X]_c \neq \emptyset$ *(i.e., X is a non-trivial BPD)*
   a) *if there is a maximal SEQUENCE-component $C \in [X]_c$, select it and goto 3d.*
   b) *if there is a well-structured component $C \in [X]_c$, select it and goto 3d.*
   c) ***if there is a minimal non-well-structured component $C \in [X]_c$, select it and***
      i. ***perform the proposed algorithm on it; if it succeeds, attach the BPEL translation to it and continue to 3e***
      ii. ***if the previous step failed, resort to the original mapping using `onEvent` construct and goto 3d.***
   d) *Attach BPEL translation of C to task object t_c.*
   e) $X := Fold(X, C, t_c)$ *and return to 3.*
4. *Output the BPEL code attached to the task object t_c.*
5. *Start event and end event are translated into a pair of `<process>` and `</process>` tags to enclose the BPEL code generated in steps 2 or 5. For terminate event, `<exit/>` activity is added.*

4 Conclusion and Future Work

In this paper we have presented an algorigthm allowing detection of a particular and very often used pattern of *partially synchronized workflow threads* in BPMN and to translate the pattern to corresponding BPEL constructs. The algorithm has been integrated into an existing generic BPMN-to-BPEL translation technique.

The key contribution of the paper is that it widens the area of workflow patterns which are nowadays possible to translate from BPMN to BPEL. The algorithm performs this without loosing the original intentions, which is, as we have pointed out in introduction, a crucial quality when automating the translation from BPMN to BPEL.

The paper thus represents a neccessary fragment in the jigsaw of the overall solution for the automated and intentions preserving BPMN-to-BPEL translation, which will eventually address the dichotomy between business process and SOA.

In the future work we would like to focus on other patterns (like multichoice workflow pattern, race condition and conditions on a sequence flow) and on verification of BPMN models (detection of deadlocks, livelocks, etc.). Also, to allow fully automated model driven development, the reverse transformation is desirable too.

Acknowledgements

This work was partially supported by the ITEA/EUREKA project OSIRIS Σ!2023.

References

1. van der Aalst, W.M.P., Lassen, K.B.: Translating Workflow Nets to BPEL4WS. In: BETA Working Paper Series, WP 145, Eindhoven University of Technology, Eindhoven (2005)
2. Diestel, R.: Graph Theory, 2nd edn. Springer, GTM 173, New York (2000)
3. Gao, Y.: BPMN-BPEL Transformation and Round Trip Engineering. eClarus Software, LLC (March 01, 2006), `http://www.eclarus.com/pdf/BPMN_BPEL_Mapping.pdf`
4. Gardner, T.: UML Modelling of Automated Business Processes with a Mapping to BPEL4WS. In: Proc. of EOOWS 2003, Darmstadt, Germany (2003)
5. Koehler, J., Hauser, R.: Untangling Unstructured Cyclic Flows — A Solution Based on Continuations. In: Proc. of CoopIS 2004, Agia Napa, Cyprus (2004)
6. Kloppmann, M., Konig, F.D., Leymann, G.F., Pfau, D.G., Roller, D.: Business Process Choreography in WebSphere: Combining the Power of BPEL and J2EE. IBM Systems Journal 43(2) (2004)
7. Lassen, K.B., van der Aalst, W.M.P.: WorkflowNet2BPEL4WS: A Tool for Translating Unstructured Workflow Processes to Readable BPEL. In: Proc. of CoopIS 2006, Montpellier, France (2006)
8. Liu, R., Kumar, A.: An analysis and taxonomy of unstructured workflows. In: Bussler, C.J., Haller, A. (eds.) BPM 2005. LNCS, vol. 3812. Springer, Heidelberg (2006)
9. Mendling, J., Lassen, K.B., Zdun, U.: Transformation strategies between block-oriented and graph-oriented process modelling languages. In: Multikonferenz Wirtschaftsinformatik 2006. Band 2. GITO-Verlag, Berlin (2006)
10. OASIS, Web Services Business Process Execution Language Version 2.0 (2005)
11. OMG, Model Driven Architecture, ormsc/01-07-01 (2001)
12. OMG, UML Superstructure Specification v2.0, formal/05-07-04 (2005)
13. Oracle, BPEL Process Manager (2007), `http://www.oracle.com/technology/products/ias/bpel/index.html`
14. Ouyang, C., van der Aalst, W.M.P., Dumas, M., ter Hofstede, A.H.M.: Translating BPMN to BPEL. Technical Report BPM-06-02, BPM center (2006)
15. Ouyang, C., Dumas, M., ter Hofstede, A.H.M., van der Aalst, W.M.P.: Pattern-based translation of BPMN process models to BPEL web services. Intl. Journal of Web Services Research (JWSR) (2007)
16. Recker, J., Mendling, J.: On the translation between BPMN and BPEL: Conceptual mismatch between process modeling languages. In: Proc. of EMMSAD 2006, Luxembourg (2006)
17. Russell, N., ter Hofstede, A.H.M., van der Aalst, W.M.P., Mulyar, N.: Workflow Control-Flow Patterns: A Revised View. BPM Center Report BPM-06-22, BPM center (2006)
18. White, S.: Using BPMN to Model a BPEL Process. BPTrends 3(3), 1, 18 (2005)
19. White, S.A.: Business Process Modeling Notation, v1.0. Business Process Management Initiative, BPMI (2004)

Quality Aware Flattening for Hierarchical Software Architecture Models

Gang Huang, Jie Yang, Yanchun Sun, and Hong Mei

Key Laboratory of High Confidence Software Technologies, Ministry of Education,
School of Electronics Engineering and Computer Science, Peking University,
Beijing, 100871, China
{huanggang, yangj, sunyc, meih}@sei.pku.edu.cn

Summary. Software architecture (SA) models play an important role in model-driven development for complex software systems. A platform-independent SA model (PIM) is usually organized as a hierarchical model via composite components for complexity control and recursive composition. However, platform-specific SA models (PSM) have to be flat because many platforms (e.g. CORBA and J2EE) do not support composite components directly. Therefore, flattening hierarchical SA models becomes a challenging step when transforming PIM to PSM. Current efforts only care about preserving the functionality of composite components during the transformation; little attention is paid to the qualities of the resulting PSM (e.g. comprehensibility, redundancy and consistency) and those of the transformation process (e.g. cost and automaticity). This paper presents a systematic approach to flattening hierarchical SA models while controlling the qualities of the resulting models and the transformation process. Our quality model can also be used as criteria to compare existing flattening approaches.

1 Introduction

In recent decades, different research topics are dedicated to ensure the qualities of complex software as well as to ease the development of complex software. Model driven development is a promising approach that emphasizes constructing software by transforming platform-independent models (PIMs) to expected platform-specific models (PSMs). The implementation can usually be generated from PSMs in an automated way. As the research of software architecture (SA) matures, SA models become good candidates of PIMs. An SA model captures the structures of a software application from a global perspective. Based on SA models, architects can perform certain analysis to see whether the application would satisfy desired qualities [19]. After an SA model is designed, the application would be developed based on the SA model. For the SA model driven development, a set of transformation rules should be designed properly in order to transform an SA model into expected PSMs.

Transformation rules vary according to specific type of implementations. In recent years, component-based implementations are popular choices. Middleware is the dominant platform for component-based systems. It provides reusable capabilities whose qualities are critical to help simplify and coordinate how components are connected and how they interoperate [18]. The transformation from platform-independent SA models to component-based PSMs is possible [15], because both models share the

R. Lee (Ed.): Soft. Eng. Research, Management & Applications, SCI 150, pp. 73–87, 2008.
springerlink.com

concept of component; however, the transformation is usually not so straightforward, because some concepts in SA are not supported properly by PSMs. Composite component (CCOM) is a typical one.

CCOM is an indispensable concept in SA [10]. It provides a black box for the design of the internal structure of a component. CCOMs may be used inside other CCOMs and thus allow hierarchical composition of complex software. This recursive definition of composition is fundamental to a component model [22]. But the concept of CCOM is not well supported by many component-based PSMs, such as those based on J2EE, CORBA and Web Services. For example, there is no "composite EJB" in J2EE [7] applications or composite CCM [4] in CORBA applications. J2EE supports the concept of "Module", which is a collection of one or more J2EE components of the same component type. However, the module is mainly for deployment purpose. A deployment module is a packaging structure for components that will ultimately execute in a run time container. No composite structures exist at runtime. Web services can form a composite service using some standards such as BPEL4WS [1] and BPML [2]. Because web services (including those constituting a new service) are global in essence, a composite service does not hide constituent services. This is quite different to the "black box" nature of CCOM in SA.

Therefore, eliminating CCOMs is inevitable during the transformation from an SA model to component-based PSMs. A component-based PSM is called a flat model, since it has no hierarchies. The elimination of CCOMs equals to the transformation from a hierarchical model to a flat one. This process is called flattening.

Some researchers have proposed approaches to flattening a hierarchical SA model [17, 9, 16, 8]. These approaches satisfy a basic requirement of flattening a hierarchical SA model, that is, the resulting PSM preserves all functions that the SA model has. Meanwhile, existing approaches pay little attention to the qualities of the resulting flat model as well as the flattening process itself. Consequently, the resulting model may be difficult to understand, and it may also be difficult to ensure the consistency between these two models.

In our research, we find that different flattening processes embody different qualities. The flattening process should be flexible enough to satisfy different quality requirements, in addition to the basic requirement of preserving functions during model transformation. In this paper, we propose a quality-aware approach to flattening a hierarchical SA model. The qualities in mind are comprehensibility, redundancy, consistency, cost, and automaticity. They are used as criteria to determine a proper flattening process. Therefore, the contribution of this approach is twofold: to guide the process of flattening a hierarchical SA model with desired qualities in mind and to compare or evaluate different flattening approaches.

The remainder is organized as follows: Section 2 briefs the hierarchical SA model and the flat PSM. Section 3 and 4 introduce our approach to flattening a single CCOM and a hierarchical SA, respectively; how this approach helps to satisfy quality requirements is given in the end of both sections. Section 5 illustrates our approach by flattening a componentized email client in different ways to satisfy different quality requirements. Section 6 discusses related work and Section 7 identifies some future work.

2 Hierarchical vs. Flat SA Model

Hierarchical SA Model

Although there is no consensus on the definition of SA, most SA models share core concepts. To make our flattening process independent of specific SA model, we use an abstract component model (Fig.1) that is adopted by most researchers. To avoid unnecessary confusion, we adopt the following convention in the context relating to SA Models hereinafter: "atomic component" refers to components that have no internal structures, "CCOM" refers to components that have internal structures, and "component" alone refers to both atomic component and CCOM.

In most SA research, an atomic component (ACOM) is modeled by interfaces and properties. An interface is either provided (e.g. P1 to Pn in Fig.1.a) or required (e.g. R1 to Rm in Fig.1.a) by its owner. What interfaces and properties an ACOM has is determined by the according component type. The concrete interface definition has little influence on a flattening process; then it is omitted in this paper.

A CCOM has interfaces and properties as an ACOM does. It is a black box component with an internal structure. The internal structure is usually treated as a sub SA (e.g., the Representation in ACME [6], or the Sub-architecture in xADL2.0 [5]). Additionally, a CCOM has mappings between its interfaces and those of internal components. Usually, an external interface is mapped into only one internal interface. The internal structure and the mappings are both defined in the according CCOM type. They are invisible to the outside of the CCOM. In this way, complicated CCOMs can be refined recursively, without causing side effects to other components.

An SA involving CCOMs presents a tree-like component hierarchy (Fig.2.a). The root is the SA for the target system, the intermediate nodes are CCOMs, and leaves are ACOMs. The link between a parent and a child node implies that the child appears inside the parent CCOM (the SA itself can be seen as a CCOM).

Flat SA Model

At the implementation level or in the platform-specific SA model, a component is a basic building block, which can be independently deployed into distributed nodes [22]. It has contractually specified interfaces and explicit context dependencies, which correspond to the provided and required interfaces of a component in SA models. The

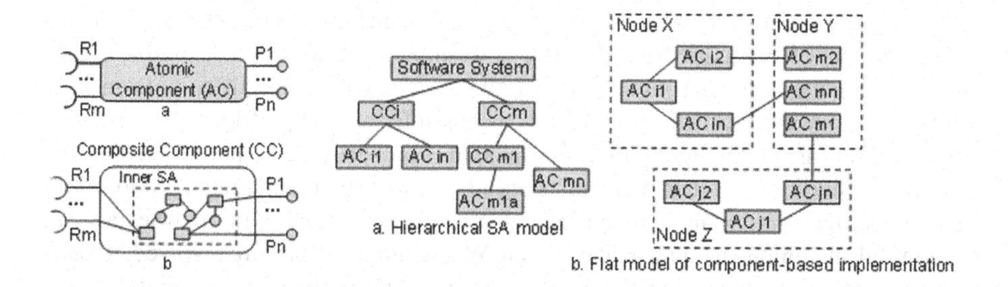

Fig. 1. ACOM and CCOM model

Fig. 2. Hierarchical and flat model

most popular component models include EJB [21], CCM [14], and COM [3]; they do not support hierarchical construction of a component directly. In other words, they support only ACOMs (Fig.1.a). Therefore, the component-based implementation model is a "flat" one (Fig.2.b). The flat model may be divided into several disjoint parts, each of which denotes an independent component execution environment. Components of the same type may be deployed into different nodes if necessary. Middleware supports interactions among components as well as security, transaction, and other common services required by a distributed application. Although these services are important, they have little influence on the flattening process. So they are omitted for brevity.

The hierarchical SA model and the flat implementation model (Fig.2) are abstractions of input and output of our flattening process. They represent most concrete SA and implementation models. Therefore, our flattening approach can be applied to SA and implementation models that conform to the abstract ones.

One key step during the transformation from a hierarchical SA model to a flat implementation model is to flattening the hierarchies; thus giving birth to a flat SA model, which is a platform-specific one, because the flattening is based on the fact that most popular component models do not support CCOM directly. Generally, the flattening is done by flattening CCOMs one by one. In the next section, a quality-aware approach to flattening a single CCOM is presented. Based on that, we will show how to flatten a hierarchical SA model from a global perspective with quality in mind.

3 Flattening a Single CCOM

To flatten a CCOM is to break the black box of CCOM and expose its internal structure. We take three steps (renaming, exposing and optimizing) to flatten a CCOM; however, different strategies adopted in each step incur different qualities. A proper composition of strategies is required by some quality requirements.

Step 1: Renaming

Before flattening, the internal structure of a CCOM is independent of the context in which the CCOM is used. Once the CCOM is flattened, the exposed internal structure may conflict with other SA elements that exist in the same context as the CCOM does. The conflict is mainly caused by name confusion, that is, when the internal structure is exposed, some elements may have identical names.

There are basically two strategies to resolve the name confusion. One is automatically renaming by predefined algorithms. For example, we can append a unique identifier to each name in the SA model, so that all involved names are different from each other. To change the original names as few as possible, we can perform renaming only when conflicts are encountered. In this case, name comparison is required (fortunately, it is easy to find out whether two names are identical). The other strategy is to notify developers to perform manual renaming when conflicts are found. In this way, a more meaningful and unique name would be given. When human efforts are involved, it tends to get a more comprehensible SA after renaming. However, human interactions hamper the automation to a certain extent.

Step 2: Exposing

In this step, the internal structure of a CCOM is exposed. To ensure that the exposing causes no loss of functionality (that is the basic flattening requirement), all links to the CCOM should be handled properly.

A straightforward way to preserve functionality is to delegate links that end with a CCOM to proper components inside the CCOM. Suppose a CCOM CC has an interface R1, which is linked to a role r of connector Cx (Fig.3.a). Inside CC, R1 is mapped to the according interface of component ComI (also named R1, which is not shown in Fig.3). The mapping implies that all connections involving R1 of CC will be delegated to that of ComI. After flattening, the role r is linked with R1 of ComI (Fig.3.b). After all links are delegated, the CC is removed.

The above link delegation preserves the functionality of a flattened CCOM; however, no information about the CCOM is saved. Therefore, it is difficult to trace back to the hierarchical model based on the flat one. As a result, if the flat model has to be modified, it may be difficult to find out the according impact on the SA model.

There are different ways to preserve the traceability between a hierarchical model and a flat one. For example, when the internal structure of a CCOM is exposed, each internal modeling element can be tagged with a special name-value pair to record where the modeling element comes from. For the above example (Fig.3.b), each modeling element (e.g. ComI) can be tagged with "!parent component, CC?" (Fig.4.a). When located deep in the component hierarchy (Fig.2.a), a component may be tagged several times. The order of these tags mirrors the order of exposing. In this way, the traceability is saved. However, these tags do not belong to the implementation model in essence. Additional facilities are required to record tags. For example, tags may be saved in deployment descriptors for EJB components in J2EE applications.

Another way to preserve traceability is to transform a CCOM into an atomic proxy component after the internal structure is exposed (Fig.4.b). In this way, no additional tags are required; instead, for each CCOM, an ACOM acting as a proxy is generated automatically. The "proxy" component plays the role of receiving all connections to a CCOM and dispatching connections to the exposed components. Therefore, the generated "proxy"

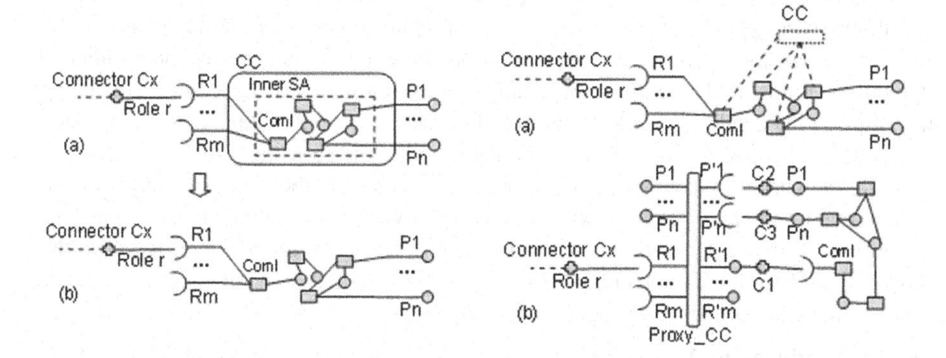

Fig. 3. Exposing without tracing information

Fig. 4. Exposing with tracing information

component not only has all interfaces a CCOM has, but also the "corresponding" interfaces of those in the CCOM. For example, the CCOM CC requires an interface R1. The "proxy" component Proxy-CC (actually, the "proxy" component can have the same name as that of the CCOM. For clarity, a prefix is added in this example) also requires an interface R1 accordingly. Besides, Proxy-CC provides a "corresponding" interface R'1 (for clarity, a different name is used. Actually, the definition of R'1 is the same as that of R1). The "corresponding" interface R'1 is used to communicate with the interface R1 in ComI. The connectors connecting the "proxy" component and the exposed ones are determined by the connectors that formerly connect to the CCOM. For example, if a procedure call connector linking to P1 uses the "pass by value" paradigm, the generated connector should also be a procedure call adopting "pass by value" paradigm; if R1 is a required message coming from a message passing connector, the generated connector should be a message passing connector, which forwards the expected message to ComI. Due to space limit, more details on connector generation are omitted. By introducing "proxy" components, the mapping defined inside CCOMs are transformed into connectors between components. The "proxy" components then serve the purpose of maintaining the traceability between a hierarchical SA model and a flat model. However, this strategy preserves traceability at the expense of relatively lower performance of the flat model, because the "proxy" components are redundant from the point of view of system functions.

Different exposing strategies imply different qualities of the implementation. Which one is more appropriate depends on which quality is predominant.

Step 3: Optimizing

The optimizing step is optional. It deserves consideration because the number of elements in a CCOM (including the root CCOM denoting the application) keeps increasing as its internal CCOMs are flattened one by one. According to the famous "7 2" principle [13], the resulting model may be too complex to understand and maintain. This step is for simplifying and revising a flat model.

Possible optimization techniques include component sharing, component type merging and splitting. For example, after a CCOM is flattened, if there are two ACOMs of the same type, and components of this type are idempotent, it is very likely that one of them can be shared, and the other can be removed. Another example, suppose there are two component types, which are semantically related. If either of them has only a few interfaces, they can be merged into a larger one. Type merging decreases the number of component types; it also creates new chances of component sharing. Many refactoring techniques [11] can be applied in this step. Due to space limit, details are not included in this paper; however, some of them are illustrated in the case study in Section 5.

It is noteworthy that most optimization techniques care about the semantics of the model; human interactions are usually necessary. Even so, we can apply some patterns for optimization, which facilitate the optimizing process to a great extent. Besides, the model optimization may corrupt the traceability links built in exposing step, unless additional measures are taken to record the changes.

Qualities Induced by Flattening Strategies. Different strategies in each step apply to the same input and produce outputs with different qualities; as a result, there are several paths to flatten a CCOM (Fig.5). Which path is more appropriate depends on expected

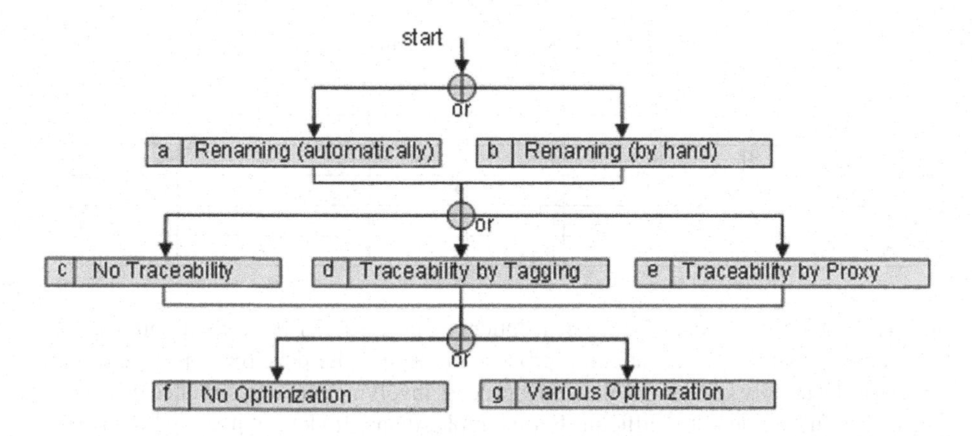

Fig. 5. Flattening framework for a CCOM

qualities. We have identified five qualities (comprehensibility, redundancy, consistency, cost and automaticity), which are either model-related (comprehensibility, redundancy, and consistency), or process-related (cost and automaticity). They are criteria to determine a specific flattening path.

- Comprehensibility: It is one important characteristic a model should have. Many techniques can be used to improve the comprehensibility of the flattening result. Renaming only when necessary as well as using meaningful names are both beneficial to enhance comprehensibility. Model optimization is also helpful because it tends to prevent a complex model from out of control.
- Redundancy: A redundant flat model means some modeling elements offer little help to implement the application's functions. Automatically generated "proxy" components are typical examples. A flat model without any optimization is more likely to have redundant modeling elements. For example, different CCOMs may have different database engines in their internal structure; however, in most cases, one database engine is sufficient. An implementation model with redundancy may consume more memories or decrease the performance of the target system, although sometimes redundancies are necessary (the "proxy" components are typical examples again). In this case, tradeoff is inevitable.
- Consistency: It means whether it is easy to ensure the flat model is consistent with the hierarchical SA model. Preserving necessary traceability information is a typical way to ensure consistency. A flattening process neglecting consistency may result in a flat model that is irrelevant to the hierarchical model. For example, adopting the strategy of exposing without tracing information will make it hard to ensure the consistency between the hierarchical SA model and the flat one.
- Cost: It includes time and human effort and should not be overwhelming when we flatten a complex SA model that has a deep component hierarchy. Lowering flattening cost often hampers other qualities. For example, the model optimization definitely cost additional time and effort. To lower the cost, no optimization or only those that require few human interactions would be adopted.

Table 1. Qualities of flattening a single CCOM

	a	b	c	d	e	f	g
Comprehensibility	-	+	-	+	+	-	+
Redundancy	0	0	+	+	-	-	+
Consistency	0	0	-	+	+	+	-
Cost	+	-	+	-	-	+	-
Automaticity	+	-	+	+	+	+	-

- Automaticity: It means whether a flattening process can be done automatically. The automaticity of a flattening process influences the development of assistant tools. Generally speaking, a flattening process involving semantics-related decisions makes full automation difficult. For example, many model optimization strategies can only be applied with the help of developers.

Although these qualities may not be complete, a table (Table 1) showing how different strategies satisfy these qualities helps to guide a flattening process. When priorities of different qualities are determined, a flattening process can be solidified by adopting strategies that benefit the predominant qualities. In Table 1, columns "a" to "g" indicate the according strategy as shown in Fig.5. The minus sign means the strategy in the column hampers the achievement of the quality of that line; the plus sign means the opposite. For example, the plus sign crossing line "Redundancy" and column "g" means the model optimization helps to reduce model redundancies. "0" means the strategy in the column is not so relevant to the according quality.

4 Flattening a Hierarchical SA Model

Technically, the approach of flattening a single CCOM is sufficient for flattening of a hierarchical SA model (Fig.2.a), which can be achieved by flattening CCOMs one by one. However, there are still some global issues relating to the aforementioned qualities. Three typical issues are discussed as follows.

Different Time for Model Separation

A component-based implementation may be deployed into several nodes; therefore, components comprising the implementation should be divided properly. One important division principle is components that have frequent interactions had better be put together [12, 20]. After every CCOM is flattened, the flat model shows clearly how ACOMs interact with each other. Based on this information, the ACOMs can be partitioned into different groups. However, for large-scale applications, it is usually time-consuming to separate a large amount of ACOMs, especially when the frequency of interactions relevant to each component seems close at first glance.

We notice that modeling elements comprising the internal structure of a CCOM usually imply relatively strong cohesion. The frequency of their interactions is relatively higher. Therefore, it may be helpful to separate the target application before all CCOMs are flattened. For example, the separation can be based on the internal structure of the root "CCOM" (i.e. the whole application). When some CCOM (say CCi) is separated into a group Gj, it means the sub tree rooted on CCi (Fig.2.a) belongs to Gj after the sub tree is flattened. This separation prior to flattening impacts the renaming step for each CCOM, because the group narrows down the scope. The cost of time for renaming may be lowered to a certain degree. However, because the separation is based on top-level coarse-grained modeling elements, the result may not be optimal.

Actually, if CCOMs are flattened in a top-down order (i.e. the top-most CCOMs are always flattened first), the similar separation can be performed when CCOMs in some level have all been flattened. This strategy is more flexible, because it combines the advantages of the former two strategies.

Top-Down vs. Bottom-Up

In fact, a hierarchical SA model can be flattened in whatever order one likes, as long as all CCOMs are flattened in the end. For example, an SA model can be flattened by flattening a CCOM randomly each time, until no CCOM is found. More naturally, because a hierarchical SA model presents a layered structure (nodes that are of the same depth in Fig.2.a form a layer), there are generally two strategies concerning the order of model flattening: top-down and bottom-up. The top-down strategy always flattens CCOMs at the highest level (when CCOMs at the highest level are all flattened, the second highest level becomes the highest one). This helps to decide when to start the separation of components, as shown in Section 4.1. The bottom-up strategy always flattens CCOMs at the lowest level (when CCOMs at the lowest level are flattened, the second lowest level becomes the lowest one). This strategy is beneficial to speed the flattening process, because some CCOMs can be flattened simultaneously, as long as the parent components of these CCOMs are different (e.g. CCik and CCm1 in Fig.2.a).

Strategy for Each CCOM

As shown in Section 3, there are several factors influencing the flattening strategies for a single CCOM. When a hierarchical SA model is to be flattened, the simplest approach is to flatten every CCOM using the same strategy. In this way, the flattening is easier to be automated. Another approach is to allow different strategies for different CCOMs. For example, to ensure consistency, most CCOMs are flattened using the "proxy" strategy. However, for CCOMs that provide functions with strict performance requirements, they may be flattened with no trace information. This approach is more flexible, but the determination of strategy for different CCOMs relies mainly on specific requirements for CCOMs as well as the relevant developers' skills. Then it is relatively difficult to automate the process.

Quality-aware Flattening of SA Models

The flattening of a hierarchical SA is based on the flattening of a single CCOM. From a global perspective, the time for model separation, the global flattening order, and the strategy for each CCOM require careful consideration. The qualities that influence the flattening of a single CCOM are also used as criteria for these issues. Different ways to handle these issues result in different flattening processes, which indicate tradeoffs among different qualities (Details are given in Table 2, which uses the same notations as those in Table 1). For example, separating the model before flattening is helpful to lower the cost of a flattening process, because it narrows down the renaming scope;

Table 2. Qualities of flattening a hierarchical SA model

	Alternatives	C_1	C_2	R	C_3	A
Model Separation	Before Flattening	+	0	-	+	+
	During Flattening	+	0	-	+	-
	After Flattening	-	0	+	-	+
Flattening Order	Randomly	0	0	0	-	+
	Top-Down	+	0	0	-	+
	Bottom-Up	+	0	0	+	+
Strategy Selection	Fixed Strategy	-	+	-	+	+
	Varied Strategy	+	-	+	-	-

Abbreviations: "C_1" for comprehensibility, "C_2" for consistency, "C_3" for cost, "R" for redundancy, "A" for automaticity.

however, it sacrifices possible optimizations for distributed components; and thus the resulting flat model may contain some redundancies.

Each combination of solutions for the three issues forms a "macro" strategy for flattening an SA model. Those introduced in Section 3 are relatively "micro" strategies. Both "macro" and "micro" strategies are equally important; they constitute our systematic approach to flattening a hierarchical SA model.

5 Illustrative Sample

To illustrate our flattening approach, different flattening processes are used to flatten a hierarchical SA model for a componentized email client (Fig.6, for space limit, snapshots of the design tool is omitted). A typical email client is capable of sending, receiving, reading, and writing emails. Several component types are identified, including Editor and EmailManagement, both of which are CCOMs (for brevity, some components, such as AddressBook and AccountManagement, are not shown in Fig.6). The internal SA models for these two CCOMs are shown in Fig.6.b and Fig.6.c. To implement the email client, the hierarchical model should be flattened.

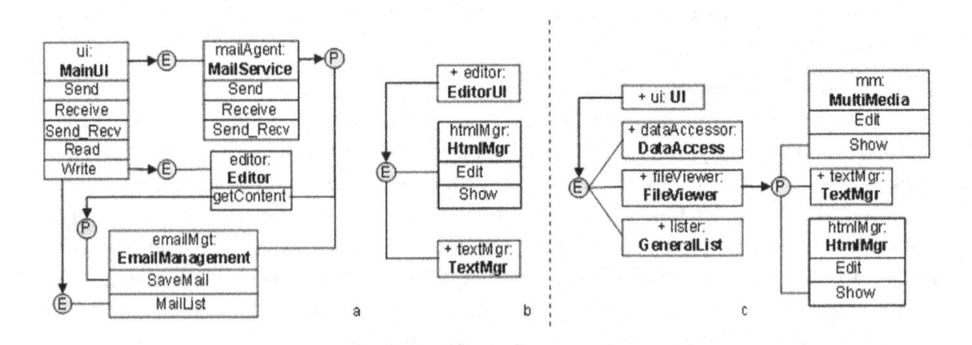

Fig. 6. SA model of an email client

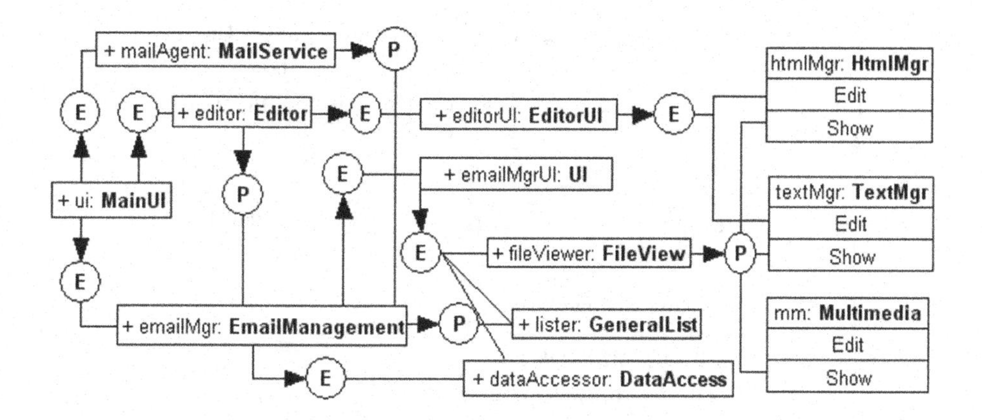

Fig. 7. Flattened email client by "b, e, g"

Assume we have only one node, so that we need not to separate the SA model. If we care about consistency and automaticity the most, a possible "macro" strategy is that all CCOMs adopt an identical "micro" strategy. Therefore, we continue to specify qualities that would drive the flattening of each CCOM. In this example, the qualities of comprehensibility and consistency are taken into account, and the former has higher priority. Therefore, a possible "micro" strategy is the composition of "b, e, g" (Fig.5), which is determined by the tool automatically.

Driven by the above quality requirements, the hierarchical model of email client would be flattened to a flat one as shown in Fig.7 (The trace information on the flattening is omitted for space limit).

- Renaming: The component editor of type EditorUI (Fig.6.b) is renamed; otherwise, it would conflict with the editor of type Editor (Fig.6.a). Another component that is renamed is the ui inside EmailManagement (Fig.6.c). By virtue of developers' knowledge on the SA model (the strategy of renaming by hand), the editor and ui are renamed to editorUI and emailMgrUI, respectively. Both new names are meaningful and self-evident.
- Exposing: After renaming, the internal structures of editor and emailMgr are exposed. Two "proxy" components, named with editor and emailMgr, are automatically generated (the interfaces are not shown in the graphical view).
- Optimization: The component htmlMgr exposed from editor and the one exposed from emailMgr (which should be renamed if CCOM editor is flattened first) are merged, that is, only one instance of HtmlMgr is needed for the display and editing of HTML format email. To ensure comprehensibility, the original name of htmlMgr remains. Another optimization is type merging and instance sharing. The component type TextMgr inside editor is different from the one inside emailMgr originally; the former is for editing text email and the latter is for displaying text email. However, these two types are lightweight and semantically related. After developers' confirmation, they are merged. After that, the two instances named textMgr originally are also merged.

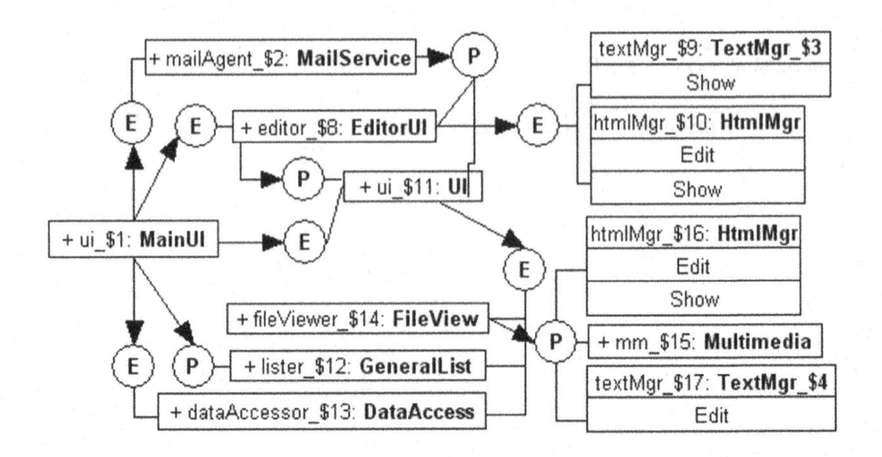

Fig. 8. Flattened email client by "a, c, f"

As quality requirements change, different flattening processes may be adopted, and thus result in different flat models. Also take the SA model of email client for example (Fig.6), if the driving qualities for the "micro" strategy are automaticity, cost and redundancy instead (their order implies the priority, that is, automaticity is of the highest priority), a possible flattening process for each CCOM would be "a, c, f" (Fig.5). The resulting flat model would then be quite different (Fig.8).

For "micro" strategy of "a, c, f" (Fig.5), the renaming is done automatically. In this case, each name is appended with a unique suffix. After exposing, no proxy components are generated. ACOMs in Fig.6 communicate directly with those exposed from editor and emailMgr. Therefore, the exposing leads to no redundancies. However, to ensure automaticity, which is of the highest priority, no optimization is applied. Consequently, two components of type HtmlMgr and two semantically relevant types (TextMgr-3 and TextMgr-4) remain in the resulting flat model. This part of the flat model could have been revised by hand to avoid redundancy; however, they remain not optimized because the automaticity is of higher priority and the optimization would decrease the automaticity. To ensure automaticity, the comprehensibility of the flat model is also sacrificed to a certain degree. By comparison, the first flat model (Fig.7) is more comprehensible than the second one (Fig.8). We also find that sometimes, it is impossible to satisfy all quality requirements, because some of them are conflict in essence. For example, automaticity is usually achieved at the cost of comprehensibility.

6 Related Work and Discussion

Rodrigues et al. proposed mapping rules from ACME to CORBA IDL [17]. No mapping rules for CCOM (Representations in ACME) are given. In our opinion, this is not a complete approach. Magee et al. proposed mappings from Darwin to IDL [9], in which CCOMs are flattened by recursive instantiation and binding delegation. In the flat model, there are no according application objects for CCOM. This approach corresponds to the "micro" strategy of "a, c, f" and the "macro" strategy of "fixed strategy". Paula et al. proposed structural mappings from ZCL to Luospace [16]. CCOMs in ZCL are transformed to glue codes, which plays the role of "proxy" components. This approach corresponds to the "micro" strategy of "a, e, f" and the "macro" strategy of "fixed strategy". Joolia et al. proposed mapping rules from ACME/ Armani to OpenCOM descriptions [8]. In this approach, a CCOM is mapped into a component framework, which is capable of extending its behavior through "plug-in" components. The naming of component frameworks comes from that of CCOMs. Therefore, this approach corresponds to the "micro" strategy of "b, e, f" and the "macro" strategy of "fixed strategy".

All the above related work cares little about the issue of model separation and flattening order. Essentially, their flattening approaches are not quality-aware. In comparison, our approach pays special attention to various quality requirements for flattening, which drive the selection of a proper flattening process.

Besides, our approach is also applicable to some special cases, to which almost all related work pays little attention.

Firstly, CCOMs usually have only simple one-to-one mappings. However, some ADLs support CCOMs with complex mappings (e.g. the mapping in ACME can be arbitrarily complex). In that case, our approach is also applicable on the whole. The step that is affected is the exposing step. In this case, exposing with "proxy" components seems more natural and appropriate. The other two strategies should be strengthened a bit to handle the complex mapping rules.

Secondly, connectors in SA are usually supported by middleware directly. For a complex connector, the flattening depends on its definition. If a complex connector is one that has an internal structure similar to that of a CCOM, the complex connector can be flattened similar to what we do to flatten a CCOM. However, if a complex connector is defined in other forms (e.g. [20]), our approach may not be applicable.

Lastly, the implementation model has no concepts that are mapped directly to CCOM. This is the most essential motivation for flattening a hierarchical SA. If some implementation model can be extended to support CCOM (e.g., Fractal, Grid Component Model, Component Array), the transformation from SA to implementation would become smoother. In that case, our approach could be applied with a simplified process. For example, the renaming could be omitted if CCOMs in SA are mapped to those at implementation one by one.

7 Future Work

There are many open issues to be addressed, including the automation of the CASE tool, more qualities to be considered, the flattening of complex connectors, and the extension of an implementation model to support composite component directly.

Acknowledgements. This effort is sponsored by the National Key Basic Research and Development Program of China (973) under Grant No. 2005CB321805; the National Natural Science Foundation of China under Grant No. 90612011; and the IBM University Joint Study Program.

References

1. BPEL4WS (2003), http://www-106.ibm.com/developerworks/webservices/library/ws-bpel
2. BPML, http://www.eclipse.org/m2m/atl/
3. COM, http://www.microsoft.com/com/
4. CORBA, http://www.corba.org/
5. Dashofy, E.M., der Hoek, A.V., Taylor, R.N.: A highly-extensible, xml-based architecture description language. In: WICSA 2001: Proceedings of the Working IEEE/IFIP Conference on Software Architecture, p. 103. IEEE Computer Society Press, Washington (2001)
6. Garlan, D., Monroe, R., Wile, D.: Acme: an architecture description interchange language. In: CASCON 1997: Proceedings of the 1997 conference of the Centre for Advanced Studies on Collaborative research, p. 7. IBM Press (1997)
7. J2EE, http://java.sun.com/javaee/
8. Joolia, A., Batista, T.V., Coulson, G., Gomes, A.T.A.: Mapping adl specifications to an efficient and reconfigurable runtime component platform. In: Fifth Working IEEE / IFIP Conference on Software Architecture (WICSA), pp. 131–140. IEEE Computer Society Press, Los Alamitos (2005)
9. Magee, J., Tseng, A., Kramer, J.: Composing distributed objects in corba. In: ISADS 1997: Proceedings of the 3rd International Symposium on Autonomous Decentralized Systems, p. 257. IEEE Computer Society Press, Washington (1997)
10. Medvidovic, N., Taylor, R.N.: A classification and comparison framework for software architecture description languages. IEEE Trans. Softw. Eng. 26(1), 70–93 (2000)
11. Mens, T., Tourwé, T.: A survey of software refactoring. IEEE Trans. Softw. Eng. 30(2), 126–139 (2004)
12. Mikic-Rakic, M., Malek, S., Beckman, N., Medvidovic, N.: A Tailorable Environment for Assessing the Quality of Deployment Architectures in Highly Distributed Settings. In: Emmerich, W., Wolf, A.L. (eds.) CD 2004. LNCS, vol. 3083. Springer, Heidelberg (2004)
13. Miller, G.: The magic number seven, plus or minus two: Some limits on our capacity for processing information. Psychological Review 63(2), 81–97 (1956)
14. OMG, CORBA Component Model Joint Revised Submission, http://www.omg.org/
15. Oreizy, P., Medvidovic, N., Taylor, R., Rosenblum, D.: Software Architecture and Component Technologies: Bridging the Gap. In: Digest of the OMGDARPA-MCC Workshop on Compositional Software Architectures, Monterey, CA (January 1998)
16. de Paula, V.C.C., Batista, T.V.: Mapping an adl to a component-based application development environment. In: Kutsche, R.-D., Weber, H. (eds.) FASE 2002. LNCS, vol. 2306, pp. 128–142. Springer, London (2002)
17. Rodrigues, M., Lucena, L., Batista, T.: From Acme to CORBA: Bridging the Gap. In: Oquendo, F., Warboys, B.C., Morrison, R. (eds.) EWSA 2004. LNCS, vol. 3047. Springer, Heidelberg (2004)
18. Schmidt, D.C., Buschmann, F.: Patterns, frameworks, and middleware: their synergistic relationships. In: ICSE 2003: Proceedings of the 25th International Conference on Software Engineering, pp. 694–704. IEEE Computer Society Press, Washington (2003)
19. Shaw, M., Clements, P.: The Golden Age of Software Architecture. IEEE Software 23(2), 31–39 (2006)

20. Spitznagel, B., Garlan, D.: A compositional approach for constructing connectors. In: WICSA 2001: Proceedings of the Working IEEE/IFIP Conference on Software Architecture, p. 148. IEEE Computer Society Press, Washington (2001)
21. Sun Microsystems Inc.: Enterprise JavaBeans Specifications, `http://www.javasoft.com`
22. Szyperski, C.: Component Software: Beyond Object-Oriented Programming. Addison-Wesley Longman Publishing Co. Inc., Boston (2002)

Case Study on Symbian OS Programming Practices in a Middleware Project

Otso Kassinen, Timo Koskela, Erkki Harjula, and Mika Ylianttila

University of Oulu, Department of Electrical and Information Engineering,
Information Processing Laboratory
P.O. BOX 4500, FIN-90014 University of Oulu, Finland
firstname.lastname@ee.oulu.fi

Summary. In this case study, we provide practical guidelines for Symbian OS software development by analyzing our experiences from a three-year mobile software research project focused on the creation of networking middleware and collaborative applications. The practices that we either conducted in the project or conceived in retrospect are summarized, and re-applicable patterns of activity in development work are identified. Based on the observations, guidelines for essential design and implementation activities are derived and potential pitfalls are addressed through example cases to provide experience-backed information for Symbian OS software developers. Among other things, the guidelines advocate the use of platform-independent solutions, the minimization of project dependencies, and the representation of complex activity sequences in a human-readable form as a routine practice.

Keywords: Symbian OS, middleware, mobile software development.

1 Introduction

Symbian OS is the leading smartphone operating system in the world, with Symbian Ltd.'s recent (January 2008) plans to spread into basic mobile phones as well. This strong position of the operating system emphasizes the importance of high-quality software development processes. Unfortunately, Symbian OS is not the most developer-friendly platform: it is regarded as a programming environment with special challenges and hard-to-learn practices for the developers [1] [2].

On such a challenging platform, wrong decisions in the software design phase may lead to significant difficulties in the implementation phase. Also incorrect implementation practices may complicate software testing and deployment. Indeed, without prior knowledge of the potential pitfalls, unforeseen problems may arise, compromising software quality or increasing development time. Moreover, specifically the creation of mobile middleware systems is a task that poses special challenges to the developers on a wireless platform [6].

In 2004 through 2007, we conducted the three-year All-IP Application Supernetworking research project, during which we designed and developed experimental mobile software for the Series 60 2nd Edition Feature Pack 2 smartphone platform, Symbian OS version 8.0. The project's main deliverables were a

R. Lee (Ed.): Soft. Eng. Research, Management & Applications, SCI 150, pp. 89–99, 2008.
springerlink.com

mobile middleware platform, called the *Plug-and-Play Application Platform* (PnPAP) [3], and collaborative applications using the middleware's services. The PnPAP middleware provides several networking-related services: the dynamic switching of different connectivities (wireless network interfaces) and peer-to-peer (P2P) networking protocols, as well as the initiation of application sessions and the distribution of context information.

Four collaborative applications were implemented on top of PnPAP; the word *collaborative* refers to the applications' ability to perform certain tasks by using the services of each other. The applications are

- *FileSharing*, a P2P file sharing application;
- *NaviP2P*, a group-interactive navigation application;
- *RealTime*, a basic mobile VoIP application; and
- *Wellness*, a sports diary for monitoring physical exercise and sharing exercise results within user groups.

In this case study, we describe our experiences from the various Symbian OS software components we developed, analyzing the observations in order to identify patterns of activity that can be repeatedly used for beneficial results. We describe what was probably done the right way and what could have been done differently. The aim of this analysis is to provide practical guidelines for developing mobile middleware and applications for Symbian OS, reducing the potential of platform-related risks realizing.

Software project management boils down to the management of complexity; each of our guidelines aims to cut down a specific type of complexity in the development process. While our work was prototype development in a research group and the number of our software-developer personnel varied between four and six, many of the guidelines should also be applicable in industry and in larger projects, where dealing with complexity is even more crucial.

The rest of this chapter is organized as follows. In Sect. 2, the timeline of the observed project is presented, and in Sect. 3, the specific experiences from the project are analyzed and practical guidelines for Symbian OS software development are derived from them. Sect. 4 summarizes the guidelines provided.

2 Timeline of the Observed Project

In this section of the case study, we provide a rough timeline of our project, to explain in what order the key software components were developed. This makes it easier to put into context the specific technical challenges analyzed later in this case study. The realized timeline of the components' development and their project-internal dependency relationships are illustrated in Fig. 1.

2.1 Project Year 2004-2005

In mid-2004, after a period of requirements gathering and initial design, we started to develop the first PnPAP version and the FileSharing application, depending on PnPAP, in parallel. The first middleware functionalities were those

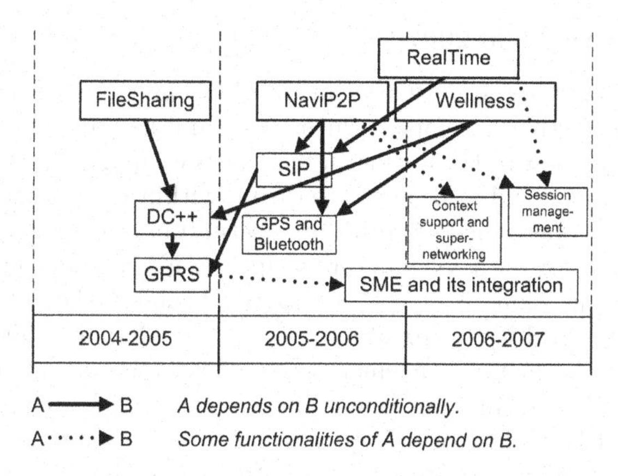

Fig. 1. Project timeline and internal dependencies

required by FileSharing: the access to P2P protocols that enable file sharing. At first, FileSharing only used a dummy protocol module that simulated P2P communication, but during spring 2005 a Direct Connect++ (DC++) file sharing protocol module was implemented so that PnPAP contained an actual P2P protocol that the FileSharing application could utilize. As PnPAP was implemented to treat protocols and connectivities as replaceable components with generic programming interfaces, also the GPRS connectivity was implemented as a module. During 2004, design work was done to identify the peer-to-peer (P2P) protocols to be implemented besides DC++. However, DC++ remained the only P2P protocol implemented during the project, as our main focus was on creating novel applications and the mechanism for the dynamic switching of modules so that the applications are always provided the best combination of a connectivity and a P2P protocol.

2.2 Project Year 2005-2006

In 2005, the first version NaviP2P application was created and was refined during winter and spring 2006; in parallel, support for location data acquisition from an external GPS device was added to PnPAP, along with a Bluetooth connectivity module. Session Initiation Protocol (SIP) was needed for PnPAP nodes' communication; in 2005, a lightweight SIP stack was created, because existing SIP stacks for Symbian did not support multi-connectivity networking. The first versions of Wellness were developed. Towards the end of this project year, there was also progress in creating a State Machine Executor (SME), to be integrated with PnPAP later. The SME's purpose was to run intelligent state machines that control the switching of connectivity and P2P protocol modules. The SME was written in Python, as opposed to all other components written in C++.

2.3 Project Year 2006-2007

In the summer of 2006, the most part of the RealTime application was created and Wellness was further developed. During the same summer and fall, PnPAP's support for application collaboration in our existing applications was extended so that applications could also be started from within each other in a context-sensitive way (this was called *application supernetworking*). Distribution of the users' personal context data within user groups was also implemented so that the context-aware applications Wellness and NaviP2P could make use of it. Wellness and NaviP2P were subjected to user tests in July-October 2006. During the fall and winter, a session management component was added to PnPAP and applied with RealTime; this component enabled the semi-automatic installation of RealTime and launching of a VoIP session in a situation, where the callee does not yet have the same application as the caller has. In the winter 2006-2007, the SME was integrated with PnPAP. PnPAP, in its final form at this project year's end, consisted of about 20.000 lines of Symbian C++ code. In addition, the applications, connectivity and protocol modules together consisted of approximately 50.000 lines of code.

3 Analysis of the Observations from the Development Work

This section of the case study elaborates the experiences from the project. Based on the experiences, practical guidelines about the operations models and choices of technologies for Symbian OS software development are presented.

3.1 Middleware Structure and API

The PnPAP core, the nerve center of the middleware within the mobile device, was designed to be a server process behind the Symbian OS Client-Server interface that is provided by the platform for Inter-Process Communication (IPC). In essence, the applications using PnPAP would issue method calls to a client-side stub Dynamic Linked Library (DLL), which implements the PnPAP Application Programming Interface (API) and translates the calls to Client-Server requests that are sent to the PnPAP core, to invoke a specific middleware behavior. On the server-side of PnPAP, every application's Client-Server connection was paired with a stub object responsible for handling that particular application's messaging. The architecture is depicted in Fig. 2; dotted lines represent process boundaries.

Early decisions (in 2004) about the internal structure and external interfaces of the PnPAP middleware affected our development efforts during the rest of the project. In retrospect, the selection of the Client-Server interface for the application-PnPAP connection was not an optimal decision. The Client-Server connection has been designed for the OS internal services with a strict "client calls, server responds" scheme. It seemed to work well for our IPC in the design phase, but when the middleware was implemented, we noticed that the lack

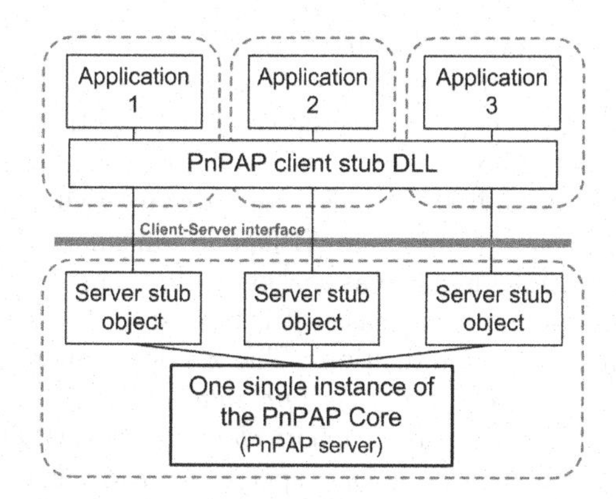

Fig. 2. Project timeline and internal dependencies

of symmetric messaging was an issue, because the applications also had to react to messages that originated from the middleware (i.e. the server calls the client). The reception of messages from the network, as well as the local application collaboration scheme, involved asynchronous messages from PnPAP to the applications.

We had to use the file system as an intermediate storage for the data in server-to-client communication. The data were written to the file by the PnPAP core; when the file was modified, the client application was signaled by a file system callback that was subscribed to with the `RFs::NotifyChange()` method. When receiving the callback event, the client would read the message from the file. Obviously this was a makeshift solution. Local-host sockets with their inherent two-way communication would have made IPC easier and more rapid to implement.

Sockets were later used in 2006-2007 for connecting PnPAP and the SME, and they proved to be more flexible than the Symbian Client-Server connection. Of course, small inconveniences such as data serialization were unavoidable in both cases, but the Symbian Client-Server interface 1) requires the programmer to learn its quirks while sockets are familiar from other platforms, 2) does not provide symmetric communication while sockets do, and 3) hinders the communication between native code and programs written in other languages. In addition, there is 4) an even more compelling reason to utilize standard solutions for central technical elements (e.g. sockets for IPC): the risk that proprietary solutions may undergo significant changes over time, which is exemplified by the following.

When we tried to port our code-base from Series 60 2nd Edition to the new 3rd Edition in 2006, the Symbian-proprietary Client-Server API, among other things, had been slightly changed in the platform. There were dozens more of similar little differences between the Editions: in the API and also in the details of project definition files, not to forget the new requirement to digitally sign the software in order to enable certain functionalities that were available in the old 2nd Edition without signing. The amalgamation of all these differences rendered the porting of the large code base a frustrating task. After two weeks of trying we gave up and decided to stay with the old 2nd Edition platform. The benefits of porting would not have been worth the costs: for us, the only actual benefit of the 3rd Edition smartphones would have been the availability of a Wireless LAN connectivity for extending PnPAP's connectivity-switching scheme.

In sharp contrast to the Symbian OS APIs, one can safely assume that the details of established technologies such as sockets will remain relatively unchanged. The less changing components, the easier it is to ensure that future software versions will work without big modifications. The key lesson here is that *if you can use a tried-and-tested, non-Symbian-specific solution, it may be a better choice than a proprietary API subject to frequent changes.* Nevertheless, it should be noted that sometimes the use of platform-specific solutions is justified. For example, dependencies to third-party software may dictate the use of Symbian-specific schemes.

3.2 Remote Node Intercommunication

To enable the PnPAP nodes to communicate with each other over the wireless networks, PnPAP offered the applications the functionality for disseminating context information and other messages embedded in SIP packets. The serialized data structure we designed for the messages was deliberately simple: an array of name-value pairs mapped to each other, values indexable by their names, with no restrictions for the values (i.e. arbitrary byte strings were accepted). Nesting of such lists, by inserting a list as a value in another list, was also supported.

Serialized data were easy to parse at the receiving node, which was a good thing, since parsing is typically more difficult than the creation of the same message, thus improvements in parsing result in more robust code. The data structure was also reused for the collaboration of local applications, which was a special case of device-internal IPC. A rule can be worded: *design the basic infrastructural components to be generic enough*, so they do not change often; also their reuse potential increases.

One key advantage in the serialization format was the inclusion of explicit length information about the contained data items, as opposed to, for example, the Extensible Markup Language (XML) in which the parser must traverse each data field until it encounters the field's ending tag. In comparison to XML, the explicit length information used in our solution improves the parsing performance of a single N-length data field from $O(N)$ to $O(1)$, makes the parser less complex, and eliminates the need for "escape characters" that protect the parser from interpreting data as special characters. As a generalized rule: *whenever you can use out-band signaling, do not use in-band signaling.* Here the word *signaling* is used in a broad sense. The rule applies to the design of protocols and file formats. Protocol design is often dependent on the other systems that are communicated with, but if easily parseable messages can be used, it may save computational resources in Symbian software.

3.3 Dynamically Pluggable Modules

Symbian's polymorphic DLLs were used for implementing the dynamically installable components (protocol and connectivity modules that can be downloaded and taken into use on-the-fly) for PnPAP. Polymorphic DLLs, by definition, implement a specific interface, and a polymorphic DLL with the same interface but different behavior can be loaded at any time, without terminating the running process. Polymorphic DLLs were a good choice for dynamically pluggable components. Besides their good on-the-fly replacement ability, polymorphic DLLs have the benefit of being able to contain arbitrary native binary code, which might seem self-evident now, but in early 2004 we had plans to implement the very P2P protocols as state machines within the SME. This would have resulted in overly complex state machines with probably quite limited ways of interacting with the operating system's services. Of course, polymorphic DLLs were far more flexible.

The state machine based solution was, in essence, an example of a task-specific language that needs to be interpreted by the actual Symbian code. The lesson learned is that *the option to implement a feature by using a task-specific language should be exercised with care.* Of course, there are also situations where the parsing of a control script or other task-specific language in Symbian software is completely justified.

3.4 Validation and Debugging with Observation of Time-Dependent Sequences

During the project, there was one specific aspect of software work that popped up again and again: the human inability to follow complex action sequences in code, either within one program or within a system of intercommunicating programs. This property of humans has been explicated in [5], where it is suggested that "even the simplest procedural logic is hard for humans to verify, but quite complex data structures are fairly easy to model and reason about. (...) Data is more tractable than program logic." We agree on this, and state that a *mapping from temporal sequences in code execution to an easy-to-follow data structure* makes software validation and debugging simpler in many cases; this is indicated by our specific experiences on the matter, described below.

In our project, it was not uncommon to be in a situation where a program error had to be searched in a sequence of 10 to 20 nested function calls, many of which were located in different source files. Additional complexity was introduced when there were multiple communicating processes or any asynchronous actions were involved. Needless to say, the complex action sequences were difficult to visualize based on the source codes and to comprehend by the developers, when trying to find out why an application crashed. This is highlighted in Symbian OS, because the ability to observe the course of program execution is very limited in a restricted device, and, moreover, the prevalent coding style of Symbian software advocates the use of complex class hierarchies that may result in long, hard-to-follow chains of method calls through several object instances, even for relatively simple actions.

Of course the debug mode in an Integrated Development Environment (IDE) allows a developer to step through running code, diminishing the need to visualize complex action sequences in one's own head when locating a misbehaving line of code. However, with our IDE and deployment platform, we could not use on-target debugging, and in practice all of our testing had to be done on target hardware (phones), which was time-consuming.

For capturing information about the temporally related actions during the execution of our Symbian software, we made heavy use of *sequence logging with plain-text log files accessible from all processes.* This simple technique proved to be a great help for resolving of both, network message flows and device-internal program sequences, as it was common to have sequences that involved call sequences through multiple processes.

One key point here is the importance of a *stateless logging facility.* Before understanding this, we used a logging code that opened a log file at the beginning of the execution and closed it only when the program exited; clearly this was stateful action. Since multiple different processes had to write to the same file, it was not smart to keep the file-handles open. The situation was fixed with stateless logging functions: when a string was to be logged, the logger function would open a file, write into it, and close it immediately. This repeated opening and closing of files caused no discernible loss of performance although Symbian software is often performance-critical. Speaking of text files, the virtues of storing information in plain-text (instead of binary structures that require dedicated access tools) is well explained in [4].

The power of visualizing complex, time-dependent chains of actions in a human-friendly way is also apparent in the sequence diagrams of the Unified Modeling Language (UML) that often facilitate the important design phase of software.

3.5 Developers' Support Functions

By developers' support functions, we mean the activities that are not strictly part of programming, but do contribute to the success or failure of a Symbian project. Symbian programming is hard enough in itself; the development team should not neglect the appropriate support functions that make their programming efforts at least manageable.

A rule that helps in daily software development work is the *assertive indication of details for the developer.* Developers should use techniques that, without much additional human effort, indicate assertively to the developer that details in the running code are as they should be and also make it easy to notice the situation when they are not.

An example of assertive indication is the use of proactively displayed build IDs or a similar solution that makes sure that the correct binary version has been updated on the handset after source-code modification. This is emphasized in Symbian OS, where the building and installation of new software versions is a multi-phase task involving multiple devices (the developer's PC and the target hardware). Our solution was to show a small indication of the binaries' successful update in a log or in a pop-up note. For example, when a particular function was altered in the sources, a once-per-build modified tag in the function's log output confirmed that the most recent version of the corresponding source file had been compiled and run. Outputting an automatically modified build ID at the beginning of the program's execution would also have been possible, but its power to confirm the source file modifications would have been less localized.

Some programmers in our team maintained a *human-readable log of their activities for their own reference*, to be prepared for cases where one suddenly needs to know, for example, how a given software module was fixed 20 days ago. Humans simply forget most of the details of solving a particular problem, but a simple text file remembers those details, for instance, the steps of advancement with the code, all in a highly readable form for a human to glance through. We recommend this kind of logging especially for programmers in Symbian OS, where hard-to-understand roundabout solutions are sometimes used and it is important to trace back how the programmer has ended up with a particular solution. Again, personal logs are an example of mapping a complex action sequence to a tractable representation.

Sometimes the Symbian OS environment seems to be restricted in ways that are not understandable. Developers should *take into account the possibly indirect consequences of these shortcomings*. For example, we have not seen Series 60 2nd or 3rd Edition phones that would display the detailed panic code on the screen (as a default action) when an application crashes. It is possible for developers to make the panic codes visible. However, if the codes were displayed as a default action, it would be beneficial for the end-user, as she could inform the software vendor about the details of a program crash. However, for some reason, the system just does not display the code; programs fail silently and often inexplicably.

This shortcoming has two corollaries: firstly, the developers must explicitly enable the indication of panics in the implementation phase; secondly, the panic codes are practically never displayed to the end-user, which is the more serious corollary of the two. This flaw in the Series 60 design causes frustration when searching the error in a program, and indirectly compromises Symbian OS software quality.

In addition to the creation of the `ErrRd` flag file that turns on the displaying of panic codes in Series 60 devices, we used the D_EXC software [7] for capturing panic codes and stack traces of crashing programs; another utility for this purpose has been introduced recently in [13], and also the Y-Tasks utility [12] works as a crash monitor, resource usage monitor and process manager.

Besides panic indication, the Series 60 2nd Edition platform lacked decent user interfaces to the running processes and to the file system. It is unclear why a smartphone does not provide a computer's basic resource-controlling tools to the user, if the phone is meant to be a handheld computer. For these functionalities, we used certain third-party utilities: TaskSpy [11] for monitoring and killing processes (Y-Tasks was not available at that time) and FExplorer [8] for accessing the file system.

3.6 Avoiding Dependencies

Dependencies to the activities of outside parties were a major cause for unexpected delays for us. In a Symbian project, it is crucial to *minimize external and internal dependencies.* If the only provider of a component fails, equivalent components may be impossible to find for the relatively exotic environment.

The most typical external dependencies in our project were the ties to parties who, for some reason, had delays in providing required software components. An example of this was our need for a software library from an outside vendor in 2004-2005. Due to the vendor's unexpected delay, the component came some 12 months late for integration with PnPAP.

Table 1. Summary of the guidelines

Guideline	Benefit for Symbian development
Using a non-platform-specific solution may often be a better choice than relying on the platform's proprietary API.	The project's dependency on the future changes of the Symbian OS API will decrease. Studying earlier Symbian OS versions reveals that changes have often been substantial.
Design the basic infrastructural components to be generic enough.	Reuse of existing code is beneficial: on Symbian OS, programming is relatively demanding and limiting the size of the codebase is favorable.
In protocols and file formats, generally use out-band signaling rather than in-band signaling.	On a mobile platform with scarce computational resources, out-band signaling makes data parsing less complex and more efficient.
Exercise the option to implement a feature by using a task-specific language only with care.	Creating an interpretation environment for a complex task-specific language may be challenging on Symbian OS.
Map temporal sequences in code execution to an easy-to-follow data structure.	Visualized sequence logs are especially useful on mobile devices with limited means of indicating the progress of a program's execution.
Use a logging facility that enables logging the actions of interrelated processes as a single sequence; access to shared logs must not involve concurrency-related issues.	See above.
Keep a human-readable log of your software development activities for your own reference.	Sometimes problems with Symbian OS force the developers to use complex solutions in the code. Personal logs may later provide information that is not available anywhere else.
Use assertive indication of details for the developer.	The developer is more confident about the details of the running code; this reduces the uncertainty and errors that may result from the complexity of Symbian software development.
Take into account the possibly indirect consequences of the shortcomings of the platform.	The behavior of Symbian OS may sometimes lead to situations that are difficult to foresee. If the problems are identified, they can often be reacted to.
Minimize your project's external and internal dependencies.	If a component provider fails, a replacement can be especially hard to find on Symbian.

Another dependency-related challenge in Symbian is the lack of standard components that are available on other platforms, decreasing the choices from which solutions can be selected. In our project, we lacked the familiar C++ Standard Template Library (STL) for Symbian OS and had to use Symbian's data-container classes instead. Penrillian [10] has recently released a STL library for Symbian OS, supporting the Leave mechanism and compatibility between descriptors and C++ strings; this is a welcome novelty. Another new tool is Nokia's Open C [9] for porting standard C libraries to Symbian OS, reducing dependencies to Symbian-specific code.

4 Conclusions

Symbian OS software developers who are facing similar challenges as those analyzed in this case study may apply the provided guidelines to keep their projects' complexity on a tolerable level. In Table 1, all the guidelines are summarized and their Symbian-specific benefits are emphasized.

References

1. Forstner, B., Lengyel, L., Kelenyi, I., Levendovszky, T., Charaf, H.: Supporting rapid application development on Symbian platform. In: Proceedings of the International Conference on Computer as a Tool. IEEE Press, Los Alamitos (2005)
2. Forstner, B., Lengyel, L., Levendovszky, T., Mezei, G., Kelenyi, I., Charaf, H.: Model-based system development for embedded mobile platforms. In: Proceedings of the Fourth and Third International Workshop on Model-Based Development of Computer-Based Systems and Model-Based Methodologies for Pervasive and Embedded Software. IEEE Computer Society Press, Los Alamitos (2006)
3. Harjula, E., Ylianttila, M., Ala-Kurikka, J., Riekki, J., Sauvola, J.: Plug-and-play application platform: Towards mobile peer-to-peer. In: Proceedings of the Third International Conference on Mobile and Ubiquitous Multimedia. ACM Press, New York (2004)
4. Hunt, A., Thomas, D.: The pragmatic programmer: From journeyman to master. Addison-Wesley, Reading (2000)
5. Raymond, E.S.: The art of Unix programming. Addison-Wesley, Reading (2003)
6. Rothkugel, S.: Towards middleware support for mobile and cellular networks: Core problems and illustrated approaches. Ph.D. thesis. University of Trier (2001)
7. Symbian D-EXC, panic code logger, `http://developer.symbian.com/` (Cited September 13, 2007)
8. Symbian FExplorer, file explorer, `http://www.gosymbian.com/` (Cited September 13, 2007)
9. Symbian Open C, `http://www.forum.nokia.com/` (Cited June 27, 2007)
10. Symbian Standard Template Library, `http://www.penrillian.com/` (Cited January 29, 2008)
11. Symbian TaskSpy, task manager, `http://www.pushl.com/taskspy/` (Cited September 13, 2007)
12. Symbian Y-Tasks, system monitor, `http://www.drjukka.com/` (Cited January 30, 2008)
13. Wang, K.: Post-mortem debug and software failure analysis on Symbian OS. M.Sc. thesis. University of Tampere (2007)

Automatic Extraction of Pre- and Postconditions from Z Specifications

Muhammad Bilal Latif and Aamer Nadeem

Center for Software Dependability
Mohammad Ali Jinnah University, Islamabad, Pakistan
m.bilal.latif@gmail.com, aamern@acm.org

Summary. Formal methods are especially useful in writing requirements specification of safety-critical systems due to their unambiguous and precise syntax and semantics. They also hold a special place in software testing phase because formal specifications provide a good opportunity for automation of the testing process. Many formal specification based software testing techniques use Finite State Machines (FSMs) because they are quite useful in test sequencing, test case generation and test case execution. Therefore, extraction of a FSM from the formal specification is an interesting problem. Although some techniques for FSM generation from formal specifications exist in the literature but none of them is fully automated. A major challenge in automatic generation of a FSM from Z specification is the identification and extraction of pre- and postconditions, since it is difficult to separate the input and output predicates when intermediate variables are used in the specification. In this paper, we present an automated approach to separate input and output predicates and extract pre- and postconditions from these predicates for a given Z operation schema. The proposed approach is supported by a tool and is also demonstrated on an example.

Keywords: FSM, specification-based testing, Z specification language.

1 Introduction

Formal methods are quite useful to model requirement specification of safety critical systems [1]. A formal specification has several advantages over specification written in a natural language. Natural language specifications are inherently ambiguous and not completely machine processable. Formal methods, on the other hand, are based on mathematical notations which make formal specifications quite clear and concise. Thus, the use of formal specification removes ambiguity, and enhances the reliability of the system.

There are many formal specification languages e.g. Z [2], Object-Z [3], VDM [4], etc. However, Z has several advantages over others. It is one of the widely used languages in industry [5]. It has been standardized by ISO (International Standards Organization) [6][7]. It is also supported by a number of automated tools [8]. Therefore, our focus in this paper is also Z language.

Combining formal specifications and software testing is an interesting and active area of research [9]. Formal specification based testing also provides significant opportunity to automate the testing process [10]. Many specification

R. Lee (Ed.): Soft. Eng. Research, Management & Applications, SCI 150, pp. 101–114, 2008.
springerlink.com

based testing techniques [11][12][13] use Finite State Machine (FSM) for modeling and testing software systems. Testing from FSM has several advantages e.g. FSMs are quite helpful in test sequencing, test case generation and test execution. Therefore, different techniques [11][14][15][13] exist in the literature for FSM generation from formal specifications. An important step in FSM generation from a formal specification is the identification of pre- and postconditions. Pre- and postconditions form the basis to determine pre- and post states of the transitions of the FSM. Identification of pre- and post conditions is also needed in partition analysis based techniques e.g. Test template framework [16] where input and output templates are generated from pre- and postconditions respectively. Most of the formal specification based testing techniques that use partition analysis are based on the classical approach of Dick and Faivre [13]. Unfortunately, Dick and Faivre demonstrated their approach on VDM specification in which pre- and postconditions are written explicitly. However, in other languages e.g. Z and Object-Z, pre- and postconditions are implicit and need to be extracted. The separation of pre- and post condition is currently a manual process and done mostly by intuition. Therefore, techniques of FSM generation from Z specification are not fully automated.

Hierons proposed an approach [17] for separating pre- and postconditions from Z operation schemas. Although the approach systematically guides towards the separation of pre- and postconditions, it does not discuss the automation issues. The work also lacked the details about how to separate input and output predicates when predicates in the specification involve intermediate variables. Our research particularly addresses these issues in detail. We propose an approach for separation of input and output predicates from Z specifications. The approach extracts the input and output predicates in three stages: elementary, transitional and secondary. Several rules are given to clearly determine the role of intermediate variables and to categorize them accordingly. We also identify certain necessary conditions for specification correctness if intermediate variables are used in the specification. Moreover, a tool is developed to support automation of the proposed approach. We also demonstrate our approach on an example.

Organization for the rest of paper is as follows. Section 2 briefly discusses the related work; section 3 presents the proposed approach, section 4 demonstrates the proposed approach on an example; section 5 discusses certain conditions that must be conformed when intermediate variables are used in the specification; section 6 proves that the proposed rules are sufficient and complete; section 7 presents the conclusion and future directions of this research.

2 Related Work

The classical work of Dick and Faivre [13] introduced the concept of generating disjoint partitions from formal specifications and then using them to identify states and transitions of FSM. This concept was then used in many other techniques. The approach was demonstrated on VDM specification in which pre- and postconditions are specified separately. However, this is not the case in other languages such as Z and Object-Z.

Hierons approach [17] is a variant of Dick and Faivre's [13] work. In Dick and Faivre's approach, VDM was used in which there was no need of identification of pre- and postconditions. Hierons focused the Z specification language and discussed the identification of pre- and post condition. However, the approach lacked certain details as discussed in section 1.

Carrington et al. [12] generated FSM from Object-Z specification of a class and used it in their approach for specification based class testing. They extended test template framework (TTF) [16] in order to generate test templates from each operation of the class under test. Then by using Murray et al's work [15] FSM was generated from test templates. Test template generation from Object-Z specification was fully automated by Ashraf and Nadeem [18] by extending an existing tool TinMan [19]. However, generation of FSM from test templates is a manual process.

Huaikou and Ling [11] purposed a framework, named as Test class framework (TCF), to formalize the process of class testing from Object-Z specification. TCF models the class with FSM for intra-class testing. The process of FSM generation resembles with that of Carrington's approach [15]. States and transitions are identified from test method templates (TMT) generated from Object-Z specification of each operation of the class under test. The notion of test method templates (TMT) was introduced by Huaikou [20] as abstract test cases for Z operation then Ling et al. [21] used them for FSM generation.

Sun and Dong [22][14] proposed an approach to extract FSM from Object-Z specifications with history invariants. The work has specially taken into account the state explosion problem of FSM generation. Other related works have not paid attention in this particular area. Predicate abstraction was effectively used to handle this problem.

3 The Proposed Approach

The approach determines the role of intermediate variables and classifies the predicates into input and output predicates. Input predicate is a predicate that defines an entry-level condition (or precondition). We denote it by P_i. Output predicate is a predicate that defines a postcondition. We denote it by P_o.

Our approach is based on two types of rules: primary rules and secondary rules. We categorize the predicates using these rules in three different stages: Elementary classification, transitional classification and secondary classification, as shown in figure 1. In the first stage, only those predicates are categorized on which primary rules are directly applicable. However, if classification has not been completed then next two stages are needed to classify the rest of predicates.

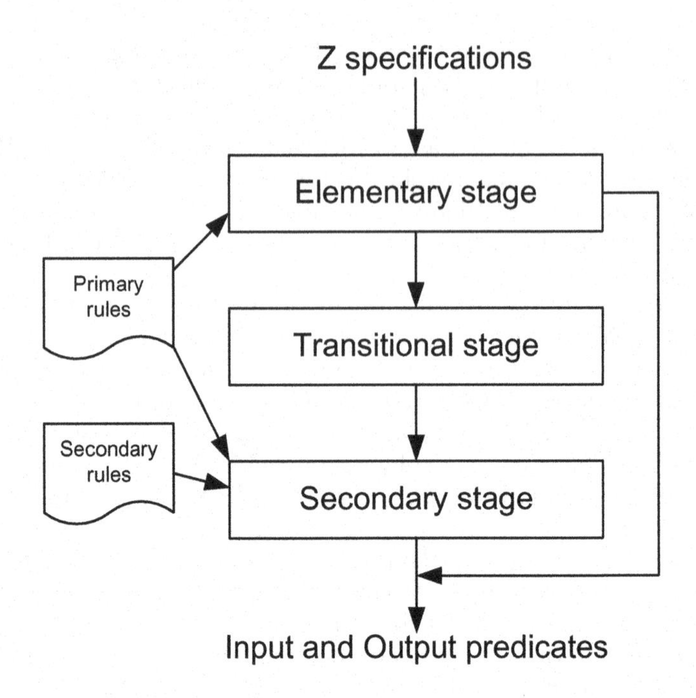

Fig. 1. Schematic diagram of the proposed approach

3.1 Input Specification

The input specification is normalized according to the following form

$$\bigwedge_{1 \leq i \leq n} X_i$$

where X is in the disjunctive normal form (DNF).

3.2 Elementary Stage

There are different kinds of variables in Z specification language i.e. input variables, output variables, before-state variables, after-state variables, and intermediate variables. We categorize input variables and before-state variables into one class named as *explicit input variables.* Similarly, output variables and after-state variables are categorized into another class named as *explicit output variables.* However intermediate variables are retained their identity yet. For convenience, we represent the predicates which involve only explicit input variables as P_{ei} and predicates which involve only explicit output variables as P_{eo}.

From the discussion of Jacky [23] on pre- and postconditions and the Hierons [17] work on input and output predicates it is clear that predicates which only involve input variables and before-state variables are input predicates and those which involve at least one output variable are output predicates. We generalize these principles in the following and call them as the primary rules.

Rule 1. The predicate which involves only those variables that belong to input variables class (IVC) is categorized as an input predicate. It involves no intermediate or output variable.

Rule 2. The predicate which involves at least one variable from output variables class (OVC) is categorized as an output predicate. It may also involve other types of variables.

Initially, only explicit input variables and explicit output variables can be added to IVC and OVC respectively. Therefore, categorization of some predicates into input and output predicates may not be possible at this stage i.e. predicates involving intermediate variables.

3.3 Transitional Stage

Now the predicates, which were not categorized in the elementary stage, are categorized into a transitional classification as follows: -

Intermediate predicate

It is a predicate which defines an intermediate variable. It may or may not involve other types of variable. An intermediate predicate, that defines an intermediate variable h, is denoted by P_h. Intermediate predicate is further categorized into two sub-types: -

Dependent intermediate predicate. An intermediate predicate which defines an intermediate variable say h in terms of at least another intermediate variable say h_1 or in terms of at least one explicit output variable is said to be a dependent P_h and is denoted by P_{hd}. The intermediate variable defined by a P_{hd} is called as dependent intermediate variable (h_d).

Independent intermediate predicate. An intermediate predicate which defines an intermediate variable without involving any other intermediate variable or an explicit output variable is said to be an independent P_h and is denoted by P_{hi}. A P_{hi} may or may not involve explicit input variables. The intermediate variable defined by a P_{hi} is called as independent intermediate variable (h_i).

Mix predicate

It is a predicate which involves explicit input variables as well as intermediate variables but no explicit output variable. It is denoted by P_{i+h}. Although mix predicate involves both input and intermediate variables like P_{hi}, it does not define an intermediate variable.

3.4 Secondary Stage

Finally, we categorize each intermediate and mix predicate as input or output predicate according to the rules presented in this section. Whenever a rule from this section is applied, IVC and OVC are updated by adding newly identified input or output variable and classification of predicates is refreshed by using primary rules.

Categorization of Independent Intermediate Predicates

Following are the rules to categorize independent intermediate variables and predicates into input and output variables and predicates respectively.

Rule 3: $P_i \wedge P_{hi}$. If there exists a predicate P_{hi} defining an intermediate variable say h_i such that P_{hi} is a conjunct with one of the input predicates P_i then P_{hi} is categorized as an output predicate and h_i is classified into output variable.

Note that this rule has no pre-requisite and is applicable wherever appropriate.

Rule 4: $P_{hi} \wedge P_{i+hi} \mid P_{o+hi} \mid P_{hd+hi}$. If there exists a predicate P_{hi} defining an independent intermediate variable h_i such that P_{hi} is a conjunct with a predicate involving the same intermediate variable as P_{hi} i.e. h_i then P_{hi} is an input predicate and h_i is classified as an input variable.

Pre-requisite: This rule is applicable only when rule 3 is not applicable.

Rule 5: $P_o \wedge P_{hi}$. If there exists an independent intermediate predicate P_{hi} defining an intermediate variable h_i such that P_{hi} is a conjunct with at least one of the output predicates P_o but not with any input predicate or predicate involving h_i then P_{hi} is output predicate and h_i is an output variable.

Rule 6: $P_{hi1} \wedge P_{hi2}$. If there exist two independent intermediate predicates P_{hi1} and P_{hi2} defining intermediate variables h_{i1} and h_{i2} such that P_{hi1} is a conjunct with P_{hi2} but neither of them is a conjunct with any input predicate or output predicate or mix predicate then $P_{hi1} \& P_{hi2}$ are input predicates and $h_{i1} \& h_{i2}$ are input variables.

Rule 7: $P_{hi1} \wedge P_{i+hi2}$. If there exists a predicate P_{hi1} defining an intermediate variable h_{i1} such that P_{hi1} is a conjunct with the mix predicate P_{i+hi2} which involves an intermediate variable other than h_{i1} say h_{i2} then first refer P_{hi2} to classify h_{i2} and finally decide about h_{i1} according to other rules.

Pre-requisite: The rule is applicable only when rule 3, 4, 5 and 6 are not applicable.

When referring to P_{hi2} then either h_{i2} can be categorized from other rules or recursive situation can be encountered as $(P_{hi2} \wedge P_{i+hi3}), (P_{hi3} \wedge P_{i+hi4}), ...,$ $(P_{hn-1} \wedge P_{i+hn})$ where 'n' is the total number of intermediate variables. If specification is correct and complete then this recursion is always solvable.

Rule 8: P_{hi}. If there exists an independent intermediate predicate P_{hi} defining an intermediate variable say h_i such that P_{hi} is not a conjunct with any other predicate then P_{hi} is categorized as an input predicate and h is classified as an input variable.

Categorization of Dependent Intermediate Predicates

For convenience, first we classify P_{hd} further into three sub-types then we give rules to categorize each sub-type into input and output predicates.

P_{hd+c} A P_{hd} which defines the dependent intermediate variable with the help of at least one output variable.

P_{hd+hi} A P_{hd} which defines the dependent intermediate variable in terms of at least one h_i but without the help of any output variable or another dependent intermediate variable.

P_{hd+hd2} A P_{hd} which defines the dependent intermediate variable h_d in terms of at least one another dependent intermediate variable say h_{d2} but without the help of any output variable.

Rule 9. P_{hd+c} is categorized as output predicate and hence h is classified as output variable.

Rule 10. Categorization of P_{hd+hi} is postponed until every independent intermediate variable h_i involved in it has been categorized. So depending upon the classification of independent intermediate variables, following are two cases: -

a. If any one h_i is categorized as output variable then rule 9 is applicable to P_{hd+hi}.
b. If all of the involved independent intermediate variables are categorized as input variables then P_{hd+hi} becomes P_{hi} and it is categorized according to the categorization rules of independent intermediate predicates.

Rule 11. Categorization of P_{hd+hd2} is postponed until the categorization of every intermediate variable involved in it except the h_d it self. Here there are two cases, which are as follows: -

a. If $P_{hd+hd2} \wedge P_{hd2}$ i.e. P_{hd+hd2} is conjunct with at least one another dependent intermediate predicate which defines h_{d2} then after deciding about all the intermediate variables involved including h_{d2}
   i. If P_{hd} involves only input variables then h_d is an input variable and P_{hd} is an input predicate.
   ii. If h_{d2} or any other intermediate variable is categorized as output variable then rule 9 is applicable.
b. If P_{hd+hd2} is not a conjunct with P_{hd2} then after deciding about all the intermediate variables involved
   i. If P_{hd} involves only input variables then h_d is re-classified as h_i.
   ii. If h_d or any other intermediate variable is categorized as output variable then rule 9 is applicable.

Categorization of Mix Predicates

Categorization of a mix predicate P_{i+h}, involving an intermediate variable h, depends on the categorization of h. If h is categorized by previous rules as input variable then P_{i+h} becomes an input predicate and if h is categorized as output variable then P_{i+h} becomes an output predicate.

3.5 Limitations

The proposed approach assumes that the input Z specification is a complete and correct specification. It must also abide by all the conditions presented in section 5 if intermediate variables have been used in the specification. Moreover, our approach is not for the derivation of implicit preconditions. Deriving implicit preconditions is a separate issue and thus is beyond the scope of this work.

4 Example

In this section, we demonstrate our approach on a simple Z specification. The example specification in figure 2 models a simple billing system for a mobile user. $a?$ is an input variable representing the air time of a user call, $pc!$ is an output variable which displays the actual cost of the call payable by user, b is a state variable keeping track of the balance of user, and t, r, c, and d are intermediate variables representing threshold call time for which promotions apply, the rate of the call per min., cost of the call without discount and discount ratio.

ManageAccount

$$a? : \mathbb{R}$$
$$pc! : \mathbb{R}$$
$$b, b' : \mathbb{R}$$

$$a? \geq 0 \land b \geq 0 \land b' \geq 0$$
$$\exists\, t, r, c, d : \mathbb{R} \bullet t = 5 \land$$
$$((a? \geq t \land r = 1.5) \lor (a? < t \land r = 2)) \land$$
$$(c = a? * r \land c \leq b) \land$$
$$((c \geq 50 \land d = 0.95) \lor (c < 50 \land d = 1)) \land$$
$$pc! = c * d \land b' = b - pc!$$

Fig. 2. Z operation schema used to demonstrate our approach

Let *Ip*, *Op*, *Iip*, *Dip*, and *Mp* be sets of input, output, independent intermediate, dependent intermediate and mix predicates respectively.
In elementary classification,

$$Ip = \{a? \geq 0, b \geq 0\}$$
$$Op = \{b' \geq 0, pc! = c * d, b' = b - ac!\}$$

In transitional classification,

$$Iip = \{t = 5, r = 1.5, r = 2, c \geq 50, c < 50, d = 0.95, d = 1\}$$
$$Dip = \{c = a? * r\}$$
$$Mp = \{a? \geq t, a? < t, c \leq b\}$$

The secondary classification is iteratively refined as explained in the following. By using rule 8, $t = 5$ is categorized input predicate and t as an input variable. Refreshing the categorization on the basis of principal rules, $a? \geq t$ and $a? < t$

are also categorized as input predicates. Now rule 3 is applicable on both $a? \geq t$ and $a? < t$, thus $r = 1.5$ and $r = 2$ are categorized as output predicates and r as output variable. Now $c = a? * r$ is categorized as output predicate and c as output variable according to rule 9. On refreshing $c = b$, $c = 50$ and $c < 50$ are categorized as output predicates. Thus the final categorization is as follows

$$
\begin{aligned}
Ip &= \{a? \geq 0, b \geq 0\ a? \geq t, a? < t\} \\
Op &= \{b' \geq 0, pc! = c * d, b' = b - ac!, r = 1.5, r = 2, c \geq 50, c < 50, \\
&\quad d = 0.95, d = 1, c \leq b, c = a? * r\}
\end{aligned}
$$

Now by using re-write rules of Hierons [17], we extract following pre- and postconditions:

$$
\begin{aligned}
&P1 : t = 5 \wedge a? \geq t \\
&Q1 : r = 1.5 \wedge c = a? * r \wedge c \leq b \wedge c \geq 50 \wedge d = 0.95 \wedge pc! = c * d \wedge \\
&\quad b' = b - ac! \wedge b' = 0 \\
&P2 : t = 5 \wedge a? \geq t \\
&Q2 : r = 1.5 \wedge c = a? * r \wedge c \leq b \wedge c < 50 \wedge d = 1 \wedge pc! = c * d \wedge \\
&\quad b' = b - ac! \wedge b' = 0 \\
&P3 : t = 5 \wedge a? < t \\
&Q3 : r = 2 \wedge c = a? * r \wedge c \leq b \wedge c \geq 50 \wedge d = 0.95 \wedge pc! = c * d \wedge \\
&\quad b' = b - ac! \wedge b' = 0 \\
&P4 : t = 5 \wedge a? < t \\
&Q4 : r = 2 \wedge c = a? * r \wedge c \leq b \wedge c < 50 \wedge d = 1 \wedge pc! = c * d \wedge \\
&\quad b' = b - ac! \wedge b' = 0
\end{aligned}
$$

In this example, P1=P2 and P3=P4 therefore it has two preconditions and four postconditions.

5 Necessary Conditions for the Usage of Intermediate Variables

The proposed approach assumes that the specification must abide by the following conditions if intermediate variables are used in the specification.

Condition I

For each intermediate variable h in the specification there must exist a P_h in the specification defining h otherwise, the specification is erroneous or incomplete.

Proof

If h has been defined in the specification then there must be a predicate defining h i.e. P_h, otherwise h becomes insignificant in the specification. This means that either the specification is incomplete or h is a redundant variable, which is always confusing.

Condition II

If $P_{h1} \wedge P_{i+h2} \mid P_{hd+h2}$ then P_{h2} cannot be a conjunct with a predicate involving h_1 where the notation "$P_{h1} \wedge P_{i+h2} \mid P_{hd+h2}$" means that P_{h1} is a conjunct with P_{i+h2} or P_{hd+h2}.
Generalizing the above condition, if $P_{h1} \wedge P_{i+h2} \mid P_{hd+h2}$ & $P_{h2} \wedge P_{i+h3} \mid P_{hd+h3}$ &...& $P_{hn-1} \wedge P_{i+hn} \mid P_{hd+hn}$ then P_{hn} cannot be a conjunct with a predicate involving h_j where $j \in 1, 2, ..., n-1$.

Proof

Let $P_{h1} \wedge P_{i+h2} \wedge P_{h2} \wedge P_{i+h1}$. It means value of h_1 depends on P_{i+h2} where as P_{i+h2} is insignificant without the value of h_2. But value of h_2 depends on P_{i+h1} which is insignificant without the value of h_1. This situation leads to an endless looping that becomes a bottleneck. Therefore, such a situation is not possible and should be avoided in the correct specification.

Corollary of condition II

For n number of intermediate variables i.e. $h_1, h_2, .., h_n$ in a specification that conforms condition II, there must be at least one P_{hk} which is not conjunct with P_{i+hl} or P_{hd+hl} where k and l are indices of intermediate variables such that $k \in \{1, 2, ..., n\}$ and $l \in \{1, 2, ..., n\} - k$.

Proof

Let every P_{hk} is conjunct with some P_{i+hl}. Also Consider a specification conforming condition II with only two intermediate variables i.e. h_1 and h_2. Here $k \in \{1, 2\}$
For $k = 1$and $l = 2, P_{h1} \wedge P_{i+h2}$
For $k = 2$ and $l = 1, P_{h2} \wedge P_{i+h1}$
This is a contradiction to condition II. Therefore, there must be at least one P_{hk} which is not conjunct with P_{i+hl}.

Condition III

If dependent intermediate variables exist in the specification then at least one h_d must be defined by P_{hd+c} or P_{hd+hi}.

Proof

Consider a specification in which dependent intermediate variables exist but none of them is defined by P_{hd+c} or P_{hd+hi}. It means all of them are defined by P_{hd+hd2}.

Further consider that specification has two dependent intermediate variables h_{d1}, h_{d2}. So h_{d1} is defined by P_{hd+hd2} and h_{d2} is defined by $P_{hd2+hd1}$ i.e. h_{d1} is defined in terms of h_{d2} and h_{d2} is defined in terms of h_{d1} which is not possible. Therefore at least one h_d must be defined by P_{hd+c} or P_{hd+hi}.

6 Completeness of Rules

In this section, we show that every possible predicate is handled and categorized by the proposed approach. This ensures that the proposed rules are complete and sufficient.

There are three major types of user declared variables in the specification i.e. input or before-state variables, output or after-state variables and intermediate variables. Any predicate P may involve one or more of these types of variables. If P involves variables of only single type then P can have three cases as follows: -

1. P involves only explicit input variables i.e. P_{ei}. It is categorized at the elementary stage.
2. P involves only explicit output variables i.e. P_{eo}. It is also categorized at the elementary stage.
3. P involves only intermediate variables i.e., intermediate predicate P_h. It is categorized at transitional and secondary stages.

If P involves variables of two types then following cases are possible: -

1. P involves only input and output variables then it is categorized by rule 2.
2. P involves only input and intermediate variables then it is a mix predicate P_{i+h}. It is categorized at secondary stage.
3. P involves only intermediate and output variables then it is categorized by rule 2.

If P involves variables of all three types then it is categorized by rule 2.

Above reasoning shows, P can syntactically fall into only four types: P_{ei}, P_{eo}, P_h, P_{i+h}. And all of those are handled successfully at appropriate stages.

It is now shown that each of the P_h and P_{i+h} is successfully categorized into input or output predicate by the proposed approach.

An intermediate variable h may be defined by a P_h as follows

1. Without using any other variable i.e. h is independent intermediate variable and defined by independent intermediate predicate P_{hi}
2. In terms of some other variable. In this case, there are following possibilities:-
   a) P_h defines h in terms of only input variables i.e. P_{hi}
   b) P_h defines h in terms of only output variables i.e. although it is P_{hd} but in this case P_h will directly categorized as output predicate by rule 3.2.
   c) defines h in terms of only intermediate variables i.e. P_{hd}
   d) P_h defines h in terms of input and output variables i.e. P_{hd} but similar situation as part b.
   e) P_h defines h in terms of input and intermediate variables i.e. P_{hd}
   f) P_h defines h in terms of only intermediate and output variables i.e. P_{hd} but similar situation as part b.

From above it is clear that the proposed approach successfully categorizes P_h into P_{hi} and P_{hd}. Now these are further categorized into input and output predications in the final categorization.

It is now shown that each intermediate and mix predicate is categorized into input and output predicate by the proposed approach.

Justification of P_{hi}

Possible scenarios for P_{hi} to occur: -

1. Lonely P_{hi}, i.e. P_{hi} is not a conjunct with any other predicate, then it is categorized according to rule 3.10.
2. P_{hi} is conjunct with
   a) P_i - rule 3 applicable
   b) P_o - rule 4 and 5 applicable
   c) P_{i+h} - rule 4 applicable
   d) P_{hi2} - rule 6 applicable
   e) P_{i+hi2} - rule 7 applicable

Justification of P_{hd}

P_{hd} may involve output variable or independent intermediate variable or dependent intermediate variable: -

1. If it involves output variable then it is categorized according to rule 9
2. If it involves independent intermediate variable then it is categorized according to rule 10
3. If it involves dependent intermediate variable then it is categorized according to rule 11.

Justification of P_{i+h}

The categorization of P_{i+h} depends upon the categorization of intermediate variable h that it involves and the categorization of h depends upon P_h defining it. Thus, the categorization of P_{i+h} depends upon the categorization of P_h defining h. It has already been shown in previous sections that the proposed approach successfully categorizes P_h and hence h. Therefore, proposed approach handles P_{i+h} smoothly.

From above discussion, it is successfully shown that the approach categorizes every possible predicate in the specification.

7 Conclusion and Future Direction

Differentiating between input and output predicates is the most important task in identifying pre- and postconditions from Z specification. Hierons gave a systematic but manual approach to separate pre- and postconditions. His approach also lacked certain details when predicates in the specification involved the intermediate variables. We addressed this issue in detail. We proposed a fully automated approach to separate input and output predicates. Our approach is also supported by a JAVA based tool. The tool takes Z specification written in LaTeX and extracts the input and output predicates. The approach was also demonstrated on an example.

The automation in specification-based testing is scarce. The identification of pre- and postconditions, which is the fundamental step in almost all of the Z based testing techniques, is currently not fully automated. Therefore, our work is an important development towards the automation of software testing techniques based on Z specifications.

The main application of identification of pre- and postconditions from Z operation is to identify pre- and post states of that operation. This is the core step in FSM extraction from specifications. Fully automated identification of pre- and post states and then automated formation of FSM from them is an important future direction of this research work.

Acknowledgments

Authors would like to thank Adnan Ashraf for his valuable feedback on early versions of this work. Muhammad Bilal Latif is supported by NESCOM Postgraduate Fellowship.

References

1. Bowen, J., Stavridou, V.: Safety-critical systems, formal methods and standards. The Software Engineering Journal 8(4), 189–209 (1992)
2. Spivey, J.M.: The Z Notation: A Reference Manual. Prentice- Hall, New York (1992)
3. Smith, G.: The Object-Z specification language. Kluwer, Boston (2000)
4. Jones, C.B.: Systematic Software Development Using VDM. Prentice-Hall, Upper Saddle River (1996)
5. Huaikou, M., Ling, L., Chuanjiang, Y., Jijun, M., Li, L.: Z User Studio: An Integrated Support Tool for Z Specifications. In: Proceedings of the Eighth Asia-Pacific Software Engineering Conference APSEC 2001, pp. 437–444. IEEE Computer Society, Washington (2001)
6. ISO/IEC, Information Technology - Z Formal Specification Notation - Syntax, Type System and Semantics. International standard ISO/IEC 13568:2002 (2002)
7. ISO/IEC, Corrigenda, Amendments and other parts of ISO/IEC 13568:2002. International standard ISO/IEC 13568:2002/Cor 1:2007 (2007)
8. Bowen, J.: The Z notation: tool support (2006), `http://vl.zuser.org/#tools` (accessed April 23, 2008)
9. Bowen, J., Bogdanov, K., Clark, J., Harman, M., Hierons, R., Krause, P.: FORTEST: Formal Methods and Testing. In: Proceedings of the 26th Annual International Computer Software and Applications Software, COMPSAC 2002, pp. 91–104. IEEE Computer Society, Washington (2002)
10. Burton, S.: Automated testing from Z specifications. Technical report, Department of Computer Science, University of York, UK (2000)
11. Huaikou, M., Ling, L.: An approach to formalizing specification-based class testing. Journal of Shanghai University 10(1), 25–32 (2006)
12. Carrington, D., MacColl, I., McDonald, J., Murray, L., Strooper, P.: From Object-Z specifications to ClassBench test suites. Software Testing, Verification and Reliability 10(2), 111–137 (2000)

13. Dick, J., Faivre, A.: Automating the generation and sequencing of test cases from model-based specifications. In: Larsen, P.G., Woodcock, J.C.P. (eds.) FME 1993. LNCS, vol. 670, pp. 268–284. Springer, Berlin (1993)
14. Sun, J., Dong, J.S.: Extracting FSMs from Object-Z specifications with history invariants. In: Proceedings of the 10th IEEE International Conference on Engineering of Complex Computer Systems ICECCS 2005, pp. 96–105. IEEE Computer Society, Washington (2005)
15. Murray, L., Carrington, D., MacColl, I., McDonald, J., Strooper, P.: Formal derivation of finite state machines for class testing. In: P. Bowen, J., Fett, A., Hinchey, M.G. (eds.) ZUM 1998. LNCS, vol. 1493, pp. 42–59. Springer, Berlin (1998)
16. Stocks, P., Carrington, D.: A framework for specification-based testing. IEEE Transactions on Software Engineering 22(11), 777–793 (1996)
17. Hierons, R.M.: Testing from a Z specification. Software Testing, Verification and Reliability 7(1), 19–33 (1997)
18. Ashraf, A., Nadeem, A.: Automating the generation of test cases from Object-Z specifications. In: Proceedings of the 30th Annual International Computer Software and Applications Software COMPSAC 2006, pp. 101–104. IEEE Computer Society, Washington (2006)
19. Murray, L., Carrington, D., MacColl, I., Strooper, P.: TinMan - A Test Derivation and Management Tool for Specification-Based Class Testing. In: Proceedings of the 32nd International Conference on Technology of Object-Oriented Languages, pp. 222–233. IEEE Computer Society, Washington (1999)
20. Huaikou, M., Ling, L.: A test class framework for generating test cases from Z specifications. In: Proceedings of the 6th IEEE International Conference on Complex Computer Systems ICECCS 2000, pp. 164–171. IEEE Computer Society, Washington (2000)
21. Ling, L., Huaikou, M., Xuede, Z.: A framework for specification-based class testing. In: Proceedings of the 8th IEEE International Conference on Complex Computer Systems, ICECCS 2002, pp. 153–162. IEEE Computer Society, Washington (2002)
22. Sun, J., Dong, J.S.: Design synthesis from interaction and state-based specifications. IEEE Transactions on Software Engineering 32(2), 349–364 (2006)
23. Jacky, J.: The way of Z: Practical Programming with Formal Methods. Cambridge University Press, New York (1996)

Towards a SPEM v2.0 Extension to Define Process Lines Variability Mechanisms

Tomás Martínez-Ruiz[1], Félix García[1], and Mario Piattini[1]

Alarcos Research Group, Information and Technology Systems Department
Escuela Superior de Informática, University of Castilla-La Mancha
Paseo de la Universidad, 4, 13071 Ciudad Real, Spain
{tomas.martinez,felix.garcia,mario.piattini}@uclm.es

Summary. Software organizations need to adapt their processes to each new project. Although Software Process Lines are the most suitable approach for the design of processes which are adapted to different contexts, SPEM does not include the appropriate mechanisms for modelling them. The objective of this paper is to suggest a SPEM extension which will support the variability implied in a Software Process Line. New variability mechanisms based on the use of Variation Points and Variants, by means of which the variability necessary in a Process Line is represented, have been proposed. The new mechanisms that we shall introduce into SPEM, will allow it to model Software Process Lines. From these lines, the generation of processes adapted to each context will simplify the selection of the appropriate variants for each variation point.

Keywords: Software Process Lines, SPEM, Variability.

1 Introduction

Software development organizations are currently interested in increasing their competitiveness and quality levels. In order to achieve this target, they need to have well-defined processes. For this reason, various process evaluation and improvement models have been proposed, such as CMMI, ISO/IEC 15504, SCAMPI, MoProSoft, EvalProSoft or the COMPETISOFT Methodological Framework processes.

However, the diversity of enterprises, projects and contexts in which processes take place is too diverse. For example, in Spain, the Civil Service Ministry requires the that totality of the software that it uses be developed using Métrica v3 [7], which is not used in the projects of other official bodies. This makes the statement *Just as there are no two identical software projects, there are no two identical software processes in the world* [5] clear. In this respect, it is difficult to apply defined generic process models in organizations without having previously adapted them to the specific situations in which they are going to be developed [18].

In the adaptation process of a software process, a set which is very similar to the original processes is developed, although these processes are not very different from each other. All these processes, which are virtually identical, make up a

R. Lee (Ed.): Soft. Eng. Research, Management & Applications, SCI 150, pp. 115–130, 2008.
springerlink.com

Software Process Family. In a process family, processes can make good use of their similarities and exploit their differences, in a manner similar to that of Software Product Lines [13].

In order to generate the differences that distinguish each process of a software process line, it is necessary to facilitate mechanisms through which this variation in the processes can be defined. However, SPEM (Software process Engineering Metamodel) [12], which has been confirmed as being the appropriate Metamodel with which to represent software processes does not, at present have the appropriate variability mechanisms to allow the generation of processes following the Software Process Line approach to take place.

In this paper, new variability mechanisms in SPEM are proposed, which contribute with the functionality necessary to model Software Process Lines. Besides this introduction, in Sect. 2 a state of the art in software product lines variability mechanisms is presented. Section 3 contains a summary of SPEM and an analysis of its current variability mechanisms, so that we can go on to show our proposal for variability mechanisms in Sect. 4. Finally, in Sect. 5 our conclusions and future work are presented.

2 State of the Art: Variability in Product Lines

To the best of our knowledge no other works exist in which the product line approach has been applied to form modelling variability in software processes, which constitutes the topic of this work.

However, some approaches dealing with variability in software product lines exist, which it is useful to analyze within the context of the work presented here.

Within the relevant literature, the majority of the variability mechanisms proposed for Software Product Lines are based on variation points and variants.

A variation point is the place in which variability occurs [2], whereas variants are the concrete elements that are placed at the variation points, each of which implements this variability in a different way.

In [17], after an analysis of the various approaches through which the variability in Software Product lines can be modelled, it was determined that the use of variants and variation points presents, amongst other advantages, the possibility of adding new variability implementations, by adding new variants. Furthermore, according to [1], it is possible to specify dependences and constraints between these elements.

In [15] a solution which is also based on variations, variation points and their relationships is proposed. This solution furthermore defines dependences between variation points and includes three abstraction levels in which to model variability. These components are included in the COVAMOF framework, which is validated by means of an experiment.

The PULSE methodology defines the five basic elements of a software product line: variation points, variants, their relations, constraints and a mechanism with which to generate products. It also includes a Decision Model in a high level of abstraction to define system variants [14].

Van der Hoeck proposes an approach based on XML schemas which model the system structural breakdown in a incremental way. One notable characteristic of this mechanism is that it allows us to manage variability owing to the evolution of versions of the product itself [4].

In [3], an idea based on a set of patterns with which to build, manage and manipulate variation points is presented.

XVLC [19] is a methodology that adds extensions to UML in order to model variation points and variants, and it furthermore specifies that the choice of a variant affects software architecture and distinguishes those variants that affect the whole system in a crosscutting manner.

Software Process Lines variability mechanisms can be designed by using the ideas from these proposals as a starting point. These mechanisms can be added to the process definition of SPEM to permit the Process Lines specification.

3 Software Process Engineering MetaModel

SPEM (Software Process Engineering Metamodel) [12] is the Object Management Group (OMG) proposal through which to represent software processes and methods. It is based on other OMG models, such as, MOF (Meta Object Facility) [10] and UML (Unified Modelling Language).

Recently the new 2.0 version of this Standard has been developed, which contributes significant new capabilities to the latest version. The Metamodel is divided into seven packages, with a hierarchical structure Fig. 1.

The following is a summary of the contents of each package illustrated in Fig. 1.

- **Core:** this contains all those classes and abstractions which are part of the metamodel base.
- **Process Structure:** this sub-package contains the elements which are necessary to define process models.
- **Process Behaviour:** this models process behaviour.
- **Managed Content:** this contains the elements which are needed to manage textual method descriptions.
- **Method Content:** this sub-package includes the concepts with which to define methods.
- **Process with Methods:** this sub-package permits method and process integration.
- **Method Plugin:** this sub-package contains the metamodel concepts with which to design and manage reusable and configurable methods and process libraries and repositories. It also includes certain variability elements and allows us to define the granularly extended process.

SPEM includes graphic notation, but uses UML association relationships. SPEM's notation allows us to represent and visualize diagrams with processes and methods.

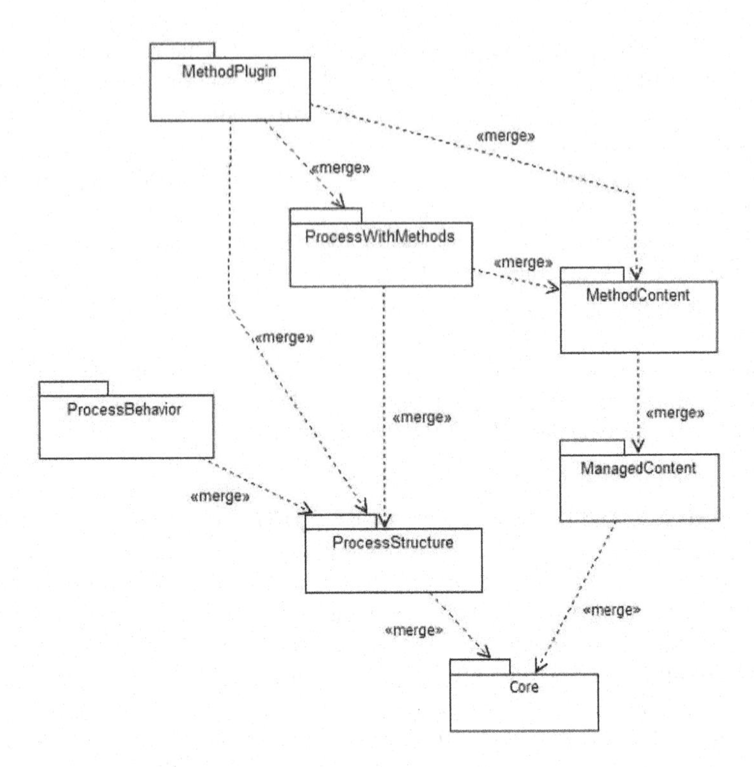

Fig. 1. Structure of the SPEM 2.0 Metamodel [12]

3.1 Variability in SPEM

SPEM v2.0 defines variability within its *MethodPlugin* package. It includes the classes which are necessary to allow variability in methods and processes, by means of defining the abstract *VariabilityElement* class and the *VariabilityType* enumeration.

The abstract *VariabilityElement* class (Fig. 2) allows us to add the variability concept to the process and the method definition of both the *ProcessStructure* and *MethodContend* packages. Given that the *Activity* (*Process Structure* package), *MethodContentElement* (*MethodContent* package) and *Section* (*ManagedContent* package) classes inherit from the VariabilityElement class, then these and their descendants acquire the "vary" capability.

By means of variation, the structure of the method can be customized without having to modify it directly.

The *VariailityType* enumeration defines the type of variability between both instances of the *VariabilityElement* class. It includes the *contributes*, *replaces*, *extends*, *extends-replaces* and *na* values (by default). The first four are now described:

Contributes is a variability relationship that allows the addition of a *VariabilityElement* to another base, without altering its original contents. This

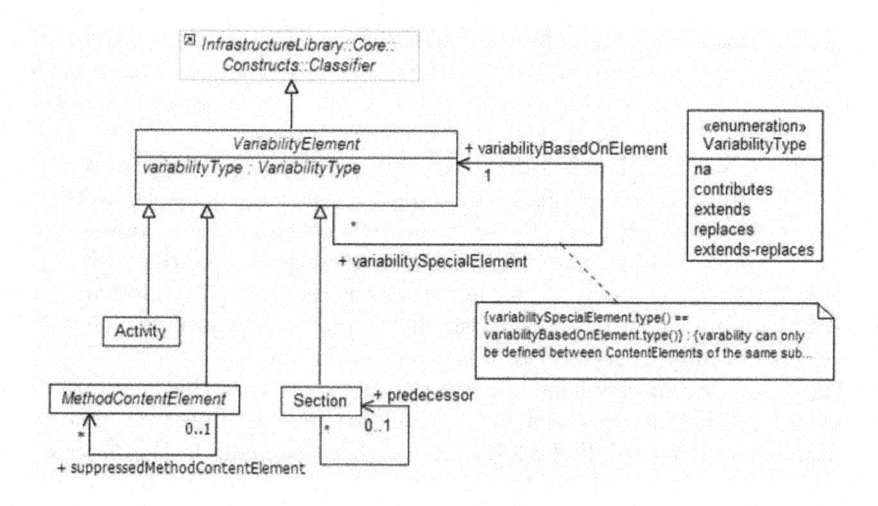

Fig. 2. *VariabilityElement* class and its relationships [12]

relationship has transitive properties. A base element must have more than one contributor.

Replaces is a variability mechanism which permits the *VariabilityElement* to be replaced by another one, without modifying its properties. A base element can only define a replaces relationship. Like contribution, the *replaces* relationship is transitive.

The **extends** relationship is an inheritance mechanism between the *VariabilityElement.* This relationship is similar to the UML *extends* relationship. This relationship is also transitive and both the contributes and replaces relationships take priority over extends.

The **extends-replaces** relationship combines the effects of both previous relationships. So while the *replace* relationship replaces all the properties of the base element, this one only replaces those values which have been redefined in the substitute element.

3.2 Limitations of the SPEM v2.0's Process Lines Variability

As a result of the analysis carried out in SPEM, certain limitations have been detected when it is used to model Software Process Lines.

1. The proposed variability in SPEM is orientated more towards changing methods than processes. In fact, the diagram in Fig. 2 shows that the *methodContent* class inherits from the *VariabilityElement* class. The methodContent class is a super-class of all the classes which permit the design of methods, and only the *Activity* class is used to model processes. Due to the fact that the processes are the entities that are executed, it will be desirable to carry out the necessary modifications/adaptations in order to execute it in the context of a project within the process, and not in the

method upon which it is based. Furthermore, we must take account that when the method content is used, the user can define the correspondence between the process and method elements, and may ignore the variability introduced into the method, so that it cannot be used.
2. This variability mechanisms do not permit the delimitation of the *core* part of the Software Process Line. As we have discussed previously, a Software Process Line is composed of a set of processes which have certain common characteristics, and which differ in other characteristics, and this makes them suitable for one context or another. However, the variability mechanisms of SPEM do not allow us to establish that a part of the process line will be common to all processes, or to limit these variations.
3. The variability mechanisms of SPEM do not guarantee that the variations introduced to generate a new process will not alter its objectives. The goals must be common to all the adjustments that are made in the process using variability mechanisms. However, because the variability mechanisms of SPEM do not allow us to define the core of a Software Process Line, it is not possible to include the objectives of the process in it and ensure that they are not modified. To avoid this, the actions of the variability mechanisms should be more limited. That is, in a process line it should be possible to specify which parts may vary and in which range of values.
4. It may be difficult to reuse SPEM v2.0 plugins with the variability mechanisms defined. In the SPEM *MethodPlugin* the *MethodPlugin* elements are defined as follows: these are extensible packages that represent a physical container for process and method packages [12]. Likewise, the *replaces* operation is used in *MethodPlugins* to replace *ContentElements* such as *RoleUse*, *TaskUse*, or *Activity* with a new variant or to change the relationships between them [12].
   This adaptation approach using plugins has the disadvantage that it is necessary to define a new plugin in order to extend that of the base which contains the process specification and adapt it to the context in which it will be enacted.
5. SPEM v2.0 does not include a specific notation with which to represent the process variability. Variability is shown by using the UML association and inheritance relationships, characterized by means of stereotypes. However, it is difficult to read some of these relations.

To sum up, due to the limitations previously listed, we believe that SPEM v2.0 variability mechanisms can be enhanced for modelling variability in Software Process Lines.

4 Specific Process Lines Variability Mechanisms

We introduce an outline of our variability mechanisms proposal with the aim of allowing SPEM v2.0 to support Software Process Lines. Likewise, we analyze their integration into the SPEM metamodel currently defined, and present an example of their use.

4.1 General Description

Our proposal is based on the mechanisms with which to manage the variability defined in Software Product Lines. Our aim is to use said mechanisms to develop others which will allow us to model variability in Software Process Lines. We list the elements that we believe to be necessary below.

1. **Variation points and variants:** in the same way as was identified in the Software Product Lines, the Variation points are the point in the process model in which the variation occurs (adapted from [14]). They will be "empty" model elements that will be "filled" with variants to allow the generation of a new process from the process line which will be adapted to the context in which they are to be developed. There will be variants of each element of the model (*VActivity*, *VWorkProductDefinition*, etc.), which can be inserted at certain variation points, and which will be dependent on the context in which the process is going to be implanted. Each variation point will have a set of associated variants and will be occupied by one or more of them.

   Due to the fact that several element types take part in a process (*RoleUse*, *WorkProductUse*, *Activity*, *TaskUse*), this distinction must also be made between the different types of variant and variation point elements.
2. **Relationships and dependencies:** Both, variation points and variants define the relationships between them. Variability is an entity which is as complex as a process and cannot be seen as a precise occurrence, but as one which will affect several part of the process structure. Dependencies will therefore allow the variations introduced into a process to be consistent. That is to say, they will ensure that if a task is carried out by a role, then both are connected.
3. **Constraints:** In addition to the relationships between the elements, we also need to take into account that constraints may appear between elements. For example, a constraint might be that a mandatory variation point must be always occupied by a variant.

4.2 New Variability Mechanisms Added to SPEM v2.0

In a way similar to that in which UML profiles extend their metamodel [6], the SPEM Metamodel can be extended to support the variability mechanism defined in the previous section. Different extensions can thus be added to the *Method Plugin* standard's package. To ensure the consistence and correct use of the new elements, OCL constraints are used [11].

First, a *LProcElement* should be designed. This is an abstraction of all the elements related to the Software Process Lines variability (Fig. 3). *LProcElement* is an abstract specification of *Classifier*, from UML packages.

The two specifications of *LProcElement* shown in Fig. 3, *VarPoint* and *Variant*, are the variation point and variant representation, respectively. Variants know the variation point for which they have been designed and which they will be able to occupy.

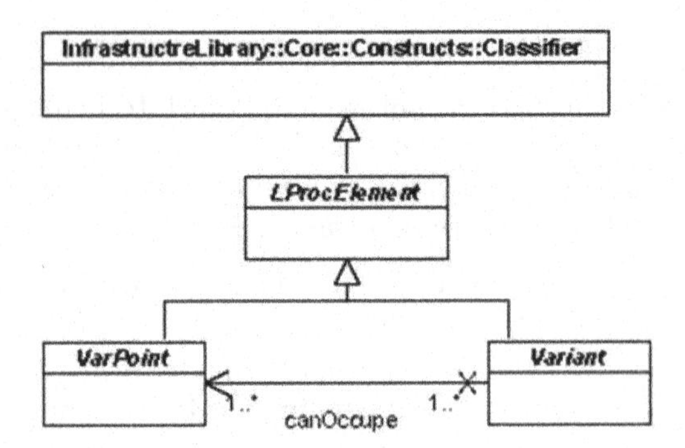

Fig. 3. Variability by means of *LProcElement*

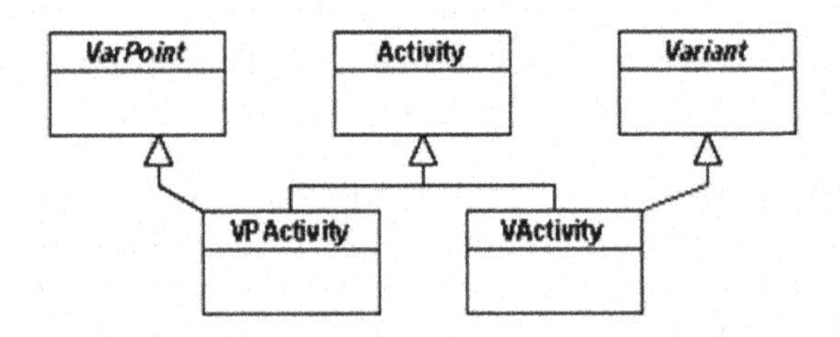

Fig. 4. *VPActivity*'s and *VActivity*'s concrete types

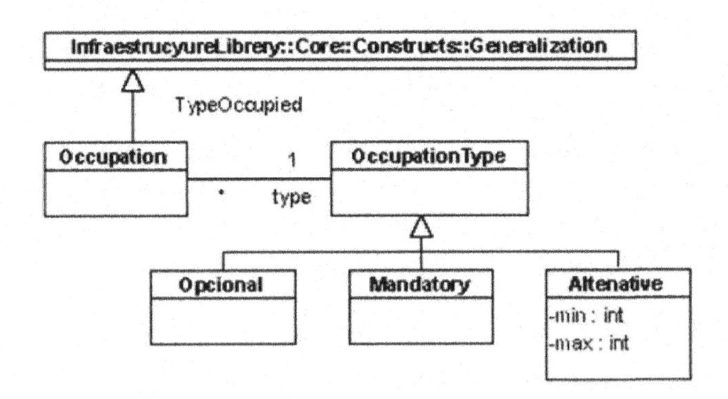

Fig. 5. *Occupation* relationship between *Variants* and *VarPoints*

The two aforementioned classes are also abstract, due to the fact that various variants and variation points of a variety of natures may exist (*Activity*, *RoleUse*). Fig. 4 shows how the variation points and variants of a concrete element were modelled. In this case, the figure shows the activity variant and variation point. As we can see, both *VPActivity* and *VActivity* are also a specification of the Activity class.

Table 1. Constraint of *Occupation* relationship between *Variation* and *Variation Point*

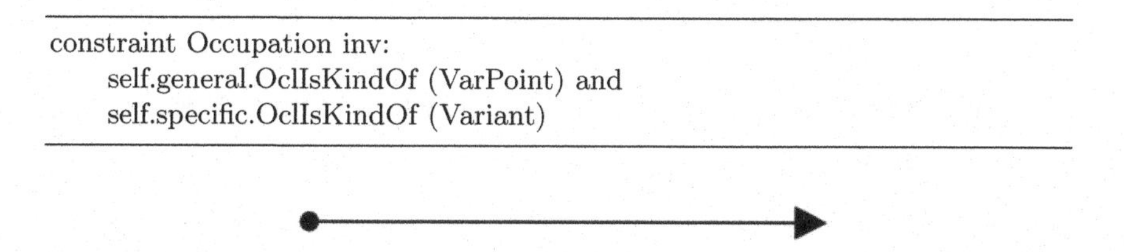

```
constraint Occupation inv:
    self.general.OclIsKindOf (VarPoint) and
    self.specific.OclIsKindOf (Variant)
```

Fig. 6. *Occupation* relationship arrow

Process models will contain different specifications of *VarPoint*, which will be "occupied" by the appropriate specifications of *Variant*. The relationship that binds a variation point to its variant, *Occupation*, is a UML Generalization specification (Fig. 5).

The *Occupation* relationship may in turn be of several types, which are specified by the *OccupationType*. These may be *Optional*, *Mandatory* or *Alternative* type. Given that this relationship is also a generalization, we need to define the OCL constraint to force it to be used exclusively between variants and variation points (Table 1).

Constraint of Table 2 specifies that the nature of the elements between which the *Occupation* relationship is established must be compatible and the variant is designed to occupy this variation point.

The *Occupation* relationship is drawn by using an arrow similar to that of the UML generalization, but with a circle on the back tip and a filled triangle on the front tip (Fig. 6), to differentiate it from the other SPEM relationships.

Finally, dependences can be established between elements, called *Variability-Dependences* (Fig. 7).

These are a UML dependencies specification, but can only exist between elements belonging to the Process Lines (Table 3).

Several types of variability dependencies may exist, such as *VariantToVariant*, *VarPointToVarPoint* or *VariantToVarPoint*, depending upon the element be-

Table 2. *Occupation* relationship between compatibility types constraint

```
constraint Occupation inv:
    (self.general.OclIsTypeOf (VPRoleDefinition) implies
    self.specific.OclIsTypeOf (VRoleDefinition)
    or self.general.OclIsTypeOf (VPActivity) implies
    self.specific.OclIsTypeOf(VActivity)
    or self.general.OclIsTypeOf (VPTaskDefinition) implies
    self.specific.OclIsTypeOf(VTaskDefinition)
    or self.general.OclIsTypeOf (VPWorkProduct) implies
    self.specific.OclIsTypeOf(VWorkProdut))
    and self.specific.canOccupe-¿includes(self.general)
```

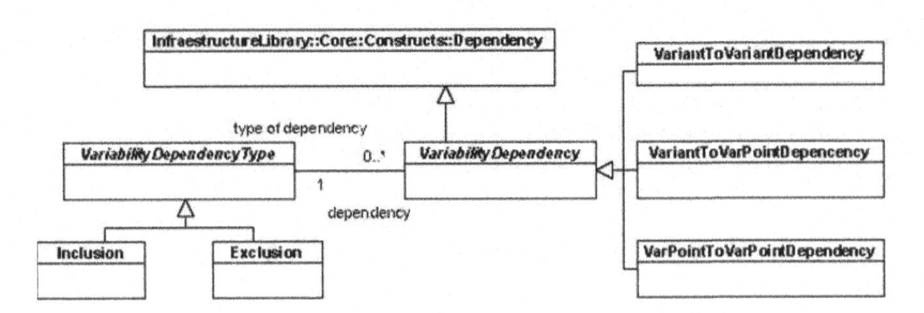

Fig. 7. *Variability Dependency* and its relationships

Table 3. *VariabilityDependency* is only available for *LProcElements*

constraint VariabilityDependency inv:
self.supplier.OclIsKindOf (LProcElement) and
self.client.OclIsKindOf (LProcElement)

Table 4. New *VarPoint* and *Variant* graphic icons

	Activity	WorkProductUse	RoleUse	TaskUse
Base Element	Activity	WorkProductUse	RoleUse	TaskUse
VarPoint	VPActivity	VPWorkProductUse	VPRoleUse	VPTaskUse
Variant	VActivity	VWorkProductUse	VRoleUse	VTaskUse

tween which they are defined. Each dependency is associated with its *Variability-DependencyType*, which identifies whether it is inclusive or exclusive. For the use of the model, we also propose new icons by means of which the new variability functionality can be graphically defined. These icons are based on those used to represent those elements to which a variation capacity is given. (Table 4).

4.3 Application Example

To illustrate our proposal, in this section we show a case study carried out on the COMPETISOFT [8, 9] process model. COMPETISOFT is an iberoamerican project in which a Software Process Improvement Framework has been

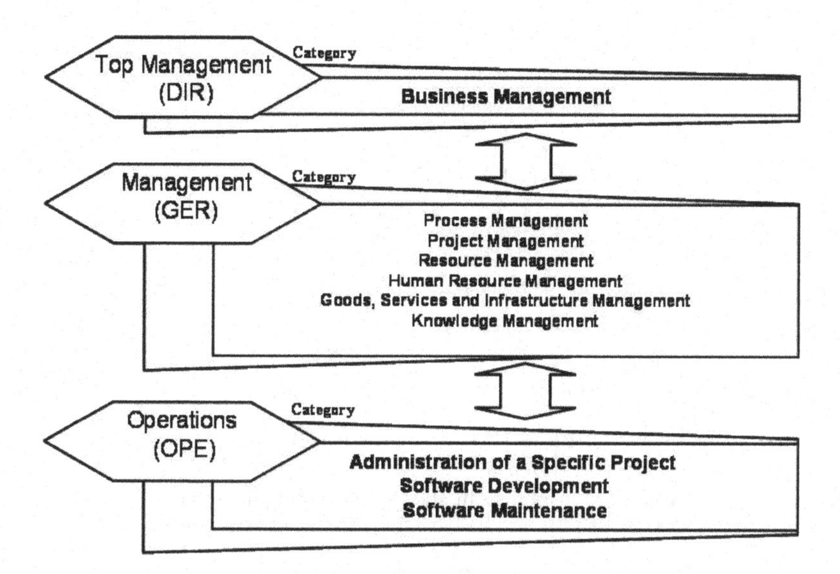

Fig. 8. Processes and categories in the COMPETISOFT process model

defined for and adapted to very small enterprises. This framework includes several process improvement proposals and is intended to be used, with the appropriate adaptations, in various iberoamerican countries. The COMPETISOFT process model is made up of ten processes which are grouped into three categories (Fig. 8). The first is Top Management and this provides the alignments which allow the organization to work. The second category, Management, includes the process, project and resource management. The Operations category deals with the practices of development projects and software maintenance.

The process model pattern includes a section called "adjustment guide" which determines possible modifications to the process which should not affect its objectives. This guide includes an introductory rough draft to the variability in the process. We shall now show how the variations in the software development process of COMPETISOFT allow said process to be modelled by using the SPEM extension described in this paper. A Process Line is thus generated, whose processes are adapted to the situations described in the adjustment guide. Without these variability mechanisms, a specific process model for each situation would be necessary.

1. Although the activities in COMPETISOFT corresponding with *Ananlysis* and *Design* have been considered seperately, in certain organisations it may be seen as expedient to fuse them, and to call the resulting activity *Anaysis and Design*. The process will thus have an activity type variation point, and a variant will take place which will contain (Fig. 9). An example which is similar to the aforementioned case occurs with the activities related to *Integration* and *Testing*.

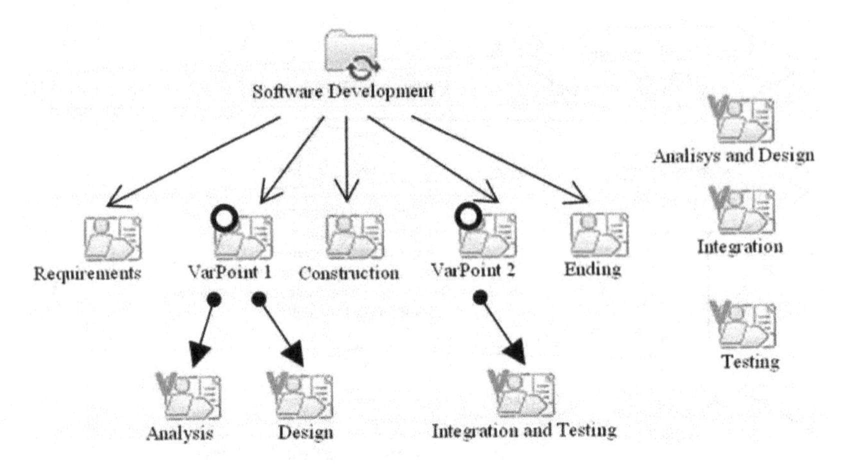

Fig. 9. Variability in the development process

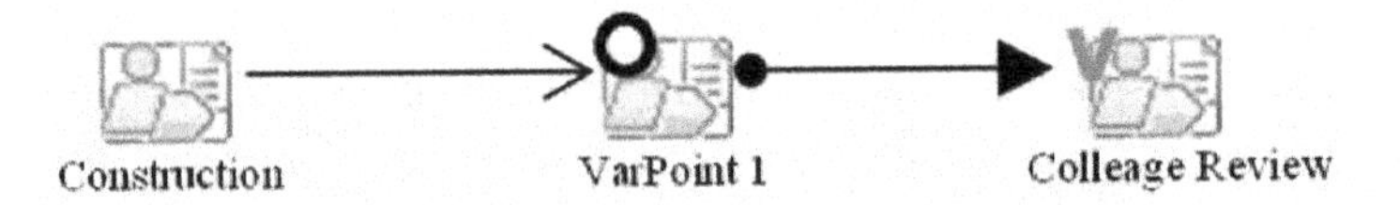

Fig. 10. Variability in optional Colleague Review

2. The *Colleague Review* (sub)activity can be added to the *Construction* activity in order to verify the code of the components. This activity will be contained within a variant, which could occupy the VarPoint found in *Construction* (Fig. 10).
3. The *Requirements Specification* work product contains various products. It may, moreover, include a *User Interface Prototype* with more or less functionality, depending upon how complicated and how important the interface is. The interface should have a variable functionality which can be modelled as a variation point and can be occupied by various WorkProduct type variants (Fig. 11).
4. The *System Testing Plan* can be validated with the Client and with the *Security System*. Such (sub)activity will thus have a role type of variation point which can be occupied by the variants which correspond with the two roles previously commented upon (Fig. 12).
5. An *IEEE Standard* can be taken into account in the *Unit Testing Definition* activity. This document will be an entry variant to the task commented upon (Fig. 13).
6. A (sub)activity of the *Modification of User Interface Prototype* can be added to the *Construction* activity, in which the User, an Expert or individuals must participate. The *Construction* activity will thus have a variation point into which the variant corresponding with the aforementioned (sub)activity

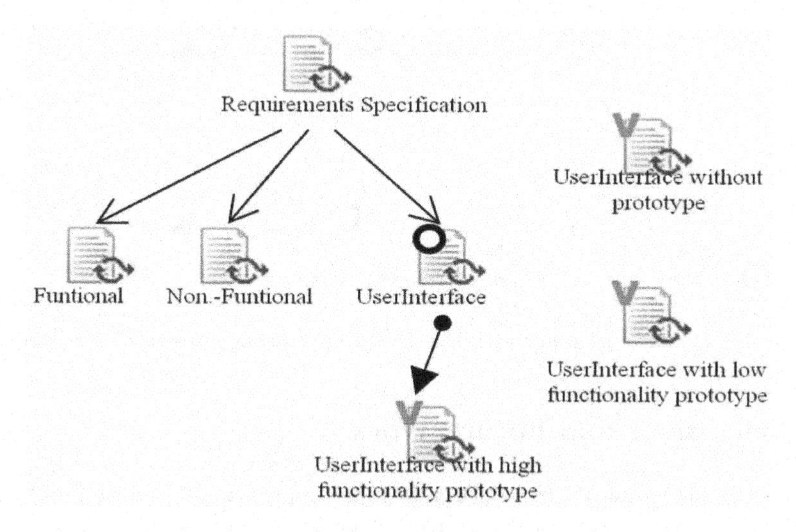

Fig. 11. Variability in the Requeriments Specification

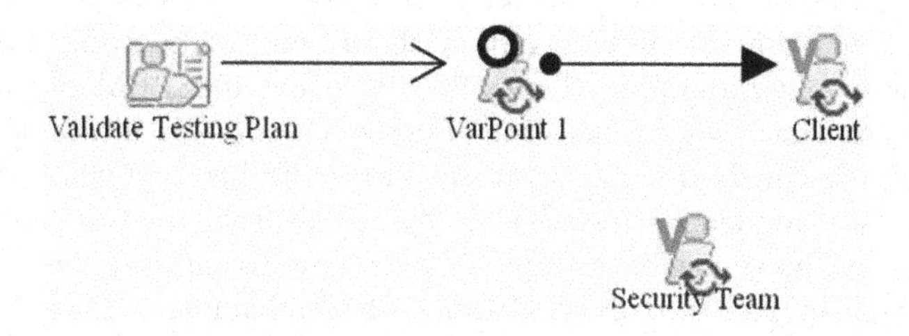

Fig. 12. Variability in the Testing Plan validation

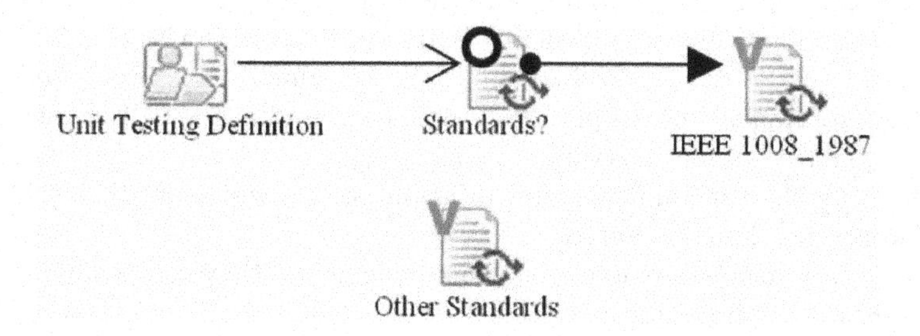

Fig. 13. Variability in the use of Standards

can be inserted. This variant will also have a variation point which can be occupied by any of the previously described roles (Fig. 14).

As this example shows, the use of the newly defined variability mechanism will allow the COMPETISOFT software development process to be adapted to various contexts. Several almost identical processes which form a part of the Software Process Lines have been generated through this adaptation.

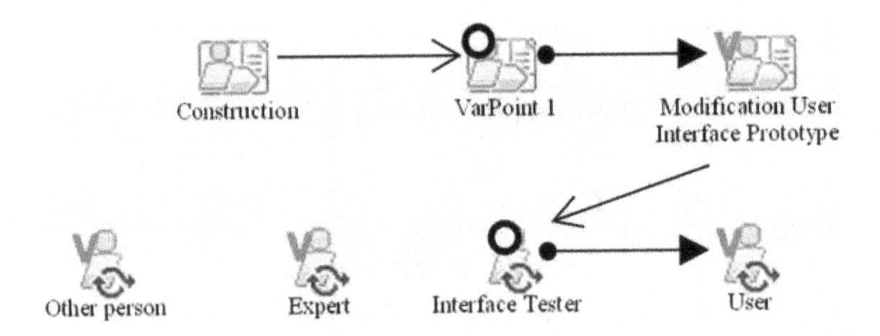

Fig. 14. Double variability with the *Modification of User Interface Prototype*

5 Conclusions and Future Work

As this work has shown, Software Process Lines approach is a powerful tool which permits the adaptation of processes to the specific conditions of an organisation and to each of the projects that it is carrying out at any given time. SPEM furthermore supposes a powerful initiative through which to model the processes which an organisation is carrying out, with the end of knowing them better, purifying them and improving them. The inclusion in SPEM of mechanisms which are appropriate to the generation of process families facilitates the creation of process models which are adapted to the particular needs of whatever is being carried out. These models are instantiated by means of the proposed mechanisms.

The approach presented in this paper allows us to manage variability from the perspective of Software Process Lines. The advantages that it offers are summarized as follows:

- This proposal is intended to measure the introduction of variability in SPEM processes, thus allowing it to define Software Process Lines.
- As with the Software Product Lines upon which they are based, these new variation mechanisms permit specification within the SPEM metamodel whose concrete parts may vary.
- By limiting the variability of the variation points, we assure that the rest of the process cannot be altered.
- Because the variants are extremely small elements, they can easily be reused at other variation points.
- This approximation includes a graphic notation with which variability can be more easily modelled in SPEM 2.0 diagrams.
- The proposed extension provides COMPETISOFT process model with a suitable notation in which the variants defined in accordance with the adjustment guide can be represented in an explicit manner.

As future work, we shall first define variability mechanisms for other constructors of the SPEM metamodel as, owing to the scope of the present work, only the core elements were considered (activities, work products, roles and tasks). Furthermore, additional empirical validation will be carried out by applying the approach to new case studies. Other future work is to automate these variability mechanisms in the EPF Composer, which supports the definition of SPEM models.

Finally, we shall also consider the aspect orientation approach to model variability. Processes are made up of determined 'aspects', such as security, usability, etc., which have a transversal effect upon all their tasks. Bearing in mind the similarity between this idea and that of the *'crosscutting concerns'*, defined in the Aspect Oriented programme [16], and the manner of interrelation which gives place to the final code, the definition of the variability mechanisms in the Software Process Lines can also be carried out by using the features of the Aspect Oriented programme.

Acknowledgement. This work is partially supported by the investigation about Software Process Lines sponsored by Sistemas Técnicos de Loterías del Estado S.A. in the framework of the aggreement about the Innovación del Entorno Metodológico de Desarrollo y Mantenimiento de Software, and by the projects ESFINGE TIN2006-15175-C05-05 financed by spanish Science and Technology Ministry and INGENIO financed by the Junta de Comunidades de Castilla-La Mancha, Consejería de Educación y Ciencia, PAC08-0154-9262.

References

1. Asikainen, T., Männistö, T., Soimimen, T.: Kumbang: A domain ontology for modeling variability in software product families. Advanced Engineering Informatics 21(1), 23–40 (2007)
2. Clauβ, M.: Generic modeling using uml extensions for variability. In: Workshop on Domain Specific Visual Languages, Tampa Bay, Florida (2001)
3. Goedicke, M., Koellmann, C., Zdun, U.: Designing runtime variation points in product line architectures: three cases. Science of Computer Programming 53(3), 353–380 (2004)
4. van der Hoeck, A.: Design-time product line architectures for any-time variability. Science of Computer Programming 53(1), 285–304 (2004)
5. Humphrey, W.S.: Managing the Software Process. Addison-Wesley, Reading (1989)
6. Korherr, B., List, B.: A UML 2 profile for variability models and their dependency to business processes. In: DEXA, Regensburg, Germany, pp. 829–834 (2007)
7. MAP. Métrica. versión 3. metodología de planificación, desarrollo y mantenimiento de sistemas de información. Ministerio de Administraciones Públicas (2005)
8. Oktaba, H., Garcia, F., Piattini, M., Pino, F., Alquicira, C., Ruiz, F.: Software process improvement in small latin-american organizations: Competisoft project. IEEE Computer 40(10), 21–28 (2007)
9. Oktaba, H., Piattini, M., Pino, F., Garcia, F., Alquicira, C., Ruiz, F., Martínez, T.: Competisoft: A improvement strategy for small latin-american software organizations. In: Oktaba, H., Piattini, M. (eds.) Software Process Improvement for Small and Medium Enterprises: Techniques and Case Studies. Idea Group Inc. (2008)
10. OMG. Meta object facility (mof) specification version 2.0. Technical report, Object Management Group (2004)
11. OMG. The object constraint language specification- version 2.0. Technical report, Object Management Group (April 2004)

12. OMG. Software process engineering metamodel specification. Technical Report ptc/07-03-03, Object Management Group (October 2007)
13. Rombach, D.: Integrated software process and product lines. In: Li, M., Boehm, B., Osterweil, L.J. (eds.) SPW 2005. LNCS, vol. 3840, pp. 83–90. Springer, Heidelberg (2006)
14. Schmid, K., John, I.: A customizable approach to full lifecycle variability management. Science of Computer Programming 53(3), 259–284 (2004)
15. Sinnema, M., Deelstra, S.: Industrial validation of covamof. Journal of Systems and Software 49(1), 717–739 (2007)
16. Sutton, S.M.: Aspect-oriented software development and software process. In: Li, M., Boehm, B., Osterweil, L.J. (eds.) SPW 2005. LNCS, vol. 3840, pp. 177–191. Springer, Heidelberg (2006)
17. Webber, D., Gomaa, H.: Modeling variability in software product lines with the variation point model. Science of Computer Programming 53(3), 305–331 (2004)
18. Yoon, I.-C., Min, S.-Y., Bae, D.-H.: Tailoring and verifying software process. In: APSEC, Macao, pp. 202–209 (2001)
19. Zhang, H., Jarzabek, S.: XVCL: a mechanism for handling variants in software product lines. Science of Computer Programming 53(3), 381–407 (2004)

Genetic Algorithm and Variable Neighborhood Search for Point to Multipoint Routing Problem

Noor Hasnah Moin and Huda Zuhrah Ab. Halim

Institute of Mathematical Sciences
University of Malaya
50603 Kuala Lumpur
Malaysia
noor_hasnah@um.edu.my

Summary. Routing of data in a telecommunication network is important due to the vast amount of data flow in the network. Message Scheduling Problem (MSP) is an important aspect of data routing. It is the process of scheduling a set of requests where each set has a single source and multiple destinations. In this paper, we propose a hybrid Genetic Algorithm to determine the order of request to be scheduled. The proposed algorithm embeds a different Steiner Tree algorithm than those reported in [3]. The second part of the study involves designing three algorithms based on the Variable Neighborhood Search. These algorithms differ in the local search heuristics employed in searching the neighborhood. The performance of these algorithms are compared on a modified set of data taken from the OR library.

1 Introduction

Communication network management is becoming increasingly difficult due to the increasing network size, rapidly changing topology and complexity. In particular, the importance of data routing in telecommunication network can no longer be ignored due to the greater accessibility to internet and mobile communications. The need for a more reliable and efficient network is of outmost importance. Message Scheduling Problem (MSP) is the process of scheduling a set of requests where each request has a single source and multiple destinations [3]. Specifically, different requests may have different source and different destination. MSP is modeled as a Point to Multipoint Routing Problem (PMRP). PMRP solves the problem by finding the optimal routing for this set of requests. [3], [11] and [6] use a combination of Genetic Algorithm (GA) and local search technique for Steiner Tree problem to solve the PMRP.

Here, two methods, Genetic Algorithm and Variable Neighborhood Search (VNS) [7] are proposed to determine the order of the requests, and a Steiner Tree heuristic algorithm is used to route all the requests. The paper is organized as follows. The concept of Point to Multipoint routing is introduced in the following section. The concepts of GA, VNS and Steiner Tree problems are explained in Section 3, 4 and 5 respectively. Our algorithms are described in detail in Section 6 and this is followed by discussions in Section 7. Finally, conclusions are drawn in Section 8.

R. Lee (Ed.): Soft. Eng. Research, Management & Applications, SCI 150, pp. 131–140, 2008.
springerlink.com

2 Point to Multipoint Routing Problem

Most of the earlier researches concentrated on the Point to Point routing (PPRP) where each request originates from a single source and has only one destination node. However, some requests may have a single source and multiple destinations, which is common in the present telecommunication networks. This is the central feature of the Point to Multipoint Routing Problem (PMRP) [3]. We note that different requests may have different source and different destination nodes. This problem is known to be NP-hard.

Most of the earlier works in PMRP treat them as a collection of point to point request and this produces poor results and is very costly [3]. In PMRP, a single copy of the message will be sent and when the message reaches an intermediate node that is linked to two or more destination nodes or other intermediate nodes, the message will be duplicated as determined by the minimum Steiner tree found. This is in contrast with the PPRP where it is possible for multiple copies of the signal to appear on the same link or parallel links, where a considerable cost is incurred. The PMRP allows more messages to be sent through the network.

3 Genetic Algorithms

Genetic Algorithm (GA) is based on the principle of survival of the fittest, where the best chromosomes (individuals) will survive and reproduce. Basically a population that is sets of chromosomes will go through several GA operators. Each chromosome has their own fitness value and chromosomes with higher fitness value have higher chances to breed and survive into the next generation. Simple genetic algorithms have three operators; reproduction, recombination and mutations.

4 Variable Neighborhood Search

Variable Neighborhood Search method is based on exploration of a dynamic neighborhood model. It works on the principles that different neighborhoods generate different search topologies [7, 2]. Systematic change of neighborhood is applied within a local search algorithm. This local search may be applied repeatedly in order to move from the incumbent solution. The algorithm can formally be stated as follows [7]:

1. Initialization
   Select the set of neighborhood structures $Nk, k = 1,, kmax$, that will be used in the search; find an initial solution x; choose a stopping condition;
2. Repeat the following until the stopping condition is met:
   i. Set $k = 1$;
   ii. Until $k = kmax$, repeat the following steps:
      a. Shaking. Generate a point $x^{'}$ at random from the $k-th$ neighborhood $x(x' \in N_k(x))$

b. Local Search. Apply some local search method with as initial solution denote with x'' the so obtained local optimum;
c. Move or not. If this local optimum is better than the incumbent, move there ($x \leftarrow x'$), and continue the search with $N_1(k \leftarrow 1)$; otherwise, set $k \leftarrow k+1$.

There are several ways that can be used to define the neighborhood structure, for example: 1-interchange (or vertex substitution), symmetric difference between two solution, Hamming distance, vertex deletion or addition, node based or path based and k-edge exchange. In this study, we define the neighborhood k, N_k as a distance function, that is the cardinality of the symmetric difference between any two solutions V_1 and V_2 written as $\rho(V_1, V_2) = |V_1 \backslash V_2|$ or $\rho = |V_1 \triangle V_2|$ where $\triangle$ denotes the symmetric difference operator. We note that the change of neighborhood can be achieved by introducing local search heuristics as a local optimum within some neighborhood is not necessarily one within another. In this study we introduce two different local searches that emphasize on the order, swap and inversion. We refer the readers to [7] for detailed description of VNS and its variant.

5 Steiner Tree Problems in a Graph

The Steiner Tree Problems in Graph (SPG) is one of the classic problems in combinatorial optimization and is NP-complete. Given a weighted undirected graph $G = (V, E)$ consisting of a nonempty set V of v vertices and a set E, $E \subseteq V \times V$, of e edges connecting pairs of vertices. Assume that we have a set of destination nodes (including source node) V', where $V' \subseteq V$, then SPG is to find a minimum cost spanning tree or subgraph, $G' = (V_T, G_T)$, of G such that all vertices in V_T are in V' where $E_T \subseteq E$ [9].

6 Metaheuristics for Point to Multi Point Routing

The telecommunications network is modeled as a graph, where each edge or path is assigned a cost and fixed capacity. Here, we consider a special case of the PMRP where all requests have the same start time, duration, and priority. Each request is identified by the source node, destination node(s) and the capacity of the request.

6.1 Genetic Algorithm for PMRP

As in [3], this research uses an order based genetic algorithm where a chromosome is a permutation of integers and each integer represents a request number. Initially the population is a set of random permutation. The requests are routed ac-cording to the order they appear in the chromosome. To route a given request, genetic algorithm will first inspect edges with capacity less than the capacity of the request; for edges that does not have enough capacity, the cost for that particular edges are made sufficiently high (artificial cost) temporarily so that it will not be considered in the route. Then, the (near) optimal Steiner tree is determined for the request. The capacities of all the paths are then updated. The steps are repeated until all requests have been routed in the order found in the chromosome. We note that the minimum cost Steiner Tree is determined by heuristics method of Kou, Markowsky and Berman's algorithm (KMB) [10]. The KMB determines the minimum cost tree that spans all destination nodes including the source node. We adopt Dijkstra's algorithm [4] to find the shortest path in the KMB algorithm. We note that we have chosen to use Kruskal's algorithm instead of Prim's algorithm as adopted by [3] due to better time complexity where Kruskal's algorithm has a time complexity of $O(E \ log \ V)$ as opposed to Prim's algorithm of $O(V^2)$. The algorithms are described in detail in the next section.

6.2 Minimum Steiner Tree

The minimum cost Steiner Tree is solved using the heuristics method of Kou, Markowsky and Berman's algorithm (KMB) [10]. The KMB algorithm is embedded in the Genetic Algorithm and Variable Neighborhood Search. The best order is the order that route the entire request without exceeding the capacity given, with the minimum cost. The cost is calculated as the summation of the product of each request capacity and the total cost of the path used. This can be expressed mathematically as

$$\sum_{j=1}^{N} q_{rj} (\sum_{i,j \epsilon ST_{rj}} c_{ij})$$

where N is the number of requests to be routed, is the capacity of request and is the cost of using a path from node to node and this path is a member of the minimal Steiner Tree for request.

We employ the KMB [10] algorithm to determine minimal Steiner Tree. The KMB is developed to determine a tree that spans all destination nodes including the source node such that the cost is minimized. As mentioned earlier, the Steiner Tree problem embeds Dijkstra's algorithm [4] to solve the shortest path problem and Kruskal's algorithm to find the minimum spanning tree. KMB algorithm [10] can be described as follows:

INPUT: an undirected distance graph $G = (V, E, d)$ and a set of Steiner points $S \subseteq V$. d is equal to the distance of the shortest path.
OUTPUT: a Steiner tree, T_H for G and S

Step 1 Construct the complete undirected distance graph $G_1 = (V_1, E_1, d_1)$from G and S.

Step 2 Find the minimal spanning tree, T_1, of G_1 .(If there are several minimal spanning trees, pick an arbitrary one).

Step 3 Construct the subgraph, G_S, of G by replacing each edge in T_1 by its corresponding shortest path in G. (If there are shortest path, pick an arbitrary one).

Step 4 Find the minimal spanning tree, T_S of G_S. (If there are several minimal spanning trees pick an arbitrary one).
Step 5 Construct a Steiner tree, T_H, from T_S by deleting edges in T_S, if necessary so that all leaves in the T_H are Steiner points.

Steiner points S consists of destination nodes and source node.

6.3 Variable Neighborhood Search for PMRP

The solution sets for VNS is similar to GA algorithm where they represent an ordered set of requests. The neighborhood structure defines a distance function as the cardinality of the symmetric difference between any two solutionsV_1 and V_2 or E_1 and E_2 that is $\rho = |V_1 \triangle V_2|$ [8]. By using this, different neighborhood structure N_k is produced. The algorithm embeds three different types of local search that is swap, inversion and *Or*-opt. The algorithm can be formally summarized as follows:

STEP 1: Generate a random solution x.
STEP 2: Generate randomly P ($P = 10$ in this study) sets of possible solutions. Classify them according to the neighborhood of x, ($N_k(x)$).
STEP 3: Starting from $N_1(x)$, do the followings (Section 4, 2ii.b) until the stopping condition is met:
- (i) Perform local search (swap, inversion or *Or*-opt)
- (ii) Find the Steiner Tree using the KMB algorithm:
  - (a) Perform Dijkstra's algorithm to find the shortest route.
  - (b) Determine the minimum spanning tree using Kruskal's algorithm
- (iii) If this local optimum is better than x, then let x be the current local optimum and repeat STEP 3. Otherwise set $k \leftarrow k + 1$.

STEP 4: Output the best solution found.

7 Results and Discussion

For comparison purposes, we run both GA and VNS on the data set given as an example in [3]. This data set contains four messages and the characteristic of the request, that is the source node, destination nodes and the capacity of each request are tabulated in Table 1. The graphical representation of the network is presented in Figure 1. All paths (edges) have the same capacity, which is and the cost as-signed to each edge is the same as in the main example in [3].

The three main operators used in this paper are Stochastic Universal Sampling (SUS), Enhanced Edge Recombination Operator (EERO) and inversion for the se-lection, crossover and mutation respectively. The probability of crossover and mutation are 0.7 and 0.1, respectively. A population size of 20 is used for the small and medium sized problems and the maximum number of generations is fixed at 20 and 30 respectively.

Table 2 tabulates the results obtained they clearly show that our algorithms produce superior result as compared to those reported in [3]. We note that VNS1

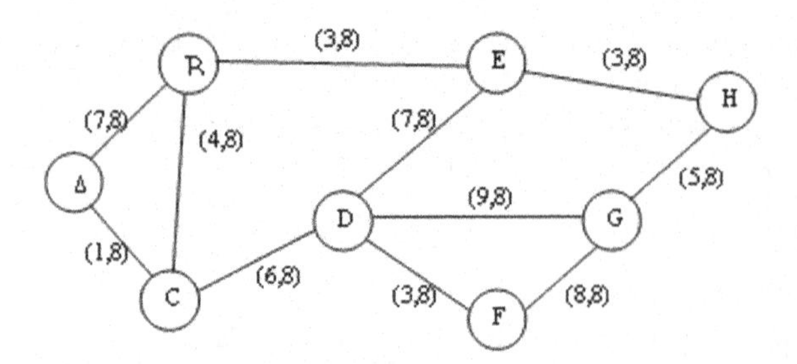

Fig. 1. A communication network with (cost, capacity) associated with each edge

Table 1. Data for the first test case

Request	Source	Destination(s)	Capacity
R1	B	C, H	5
R2	D	B, H	3
R3	G	E, F, B	4
R4	A	G	2

Table 2. Results for the first test case

	GA [3]	GA	VNS1	VNS2
Best Order	[2 4 1 3]	[2 4 1 3]	[4 2 1 3]	[1 3 2 4]
Time(in seconds)	-	1.356253	0.303125	0.257813
Cost	253	249	249	249

Table 3. Data for the second test case

Request	Source	Destination(s)	Capacity
1	36	7, 23, 25, 40	8
2	17	15, 30, 31, 40, 41, 46	5
3	48	3, 9	9
4	41	13, 22, 27, 35, 50	6
5	2	6, 14, 18, 23, 27, 33, 47, 49	5
6	13	28	6
7	50	5, 12, 28,31, 44, 45	7
8	20	17, 25, 41	8

and VNS2 embed swap and invert local search, respectively. Also GA1 and GA2 represent the results from [3] and ours, respectively. We note that our algorithms also found alternative solutions:[1 2 3 4], [1 4 3 2], [1 3 4 2], [2 1 3 4],[2 1 4 3] and [4 1 3 2] . We are not able to compare the results reported in [3] and [11] due to unavailability of data.

Table 2 illustrates the results for the example taken from [3]. It is observed that the results from our GA and the two VNS produce slightly superior results.

Table 4. Results for the second test case

No of Requests	Criteria	GA	VNS1	VNS2
4R	Best Order	[3 1 2 4]	[3 1 2 4]	[3 1 2 4]
	Time	14.21877	6.085938	6.434378
	Cost	1208	1208	1208
5R	Best Order	[5 1 3 4 2]	[5 1 3 4 2]	[5 1 3 4 2]
	Time	19.30315	10.26875	9.723444
	Cost	1703	1703	1703
6R	Best Order	[5 1 3 2 4 6]	[5 1 3 4 2 6]	[5 1 3 4 2 6]
	Time	21.65938	13.53594	11.99063
	Cost	1841	1841	1841
7R	Best Order	[5 1 3 2 4 6 7]	[5 1 2 3 4 6 7]	[5 1 3 4 2 6 7]
	Time	26.02657	14.36407	17.21718
	Cost	2450	2450	2450

Table 5. Data for the third test case

Request	Source	Destination(s)	Capacity
1	36	7, 23, 25, 40	8
2	17	15, 30, 31, 40, 41, 46	5
3	48	3, 9	9
4	41	13, 22, 27, 35, 50	6
5	2	6, 14, 18, 23, 27, 33, 47, 49	5
6	13	28	6
7	50	5, 12, 28,31, 44, 45	7
8	24	30, 29, 20	5
9	52	13, 55, 22, 9	4
10	53	28, 52, 13, 55, 41, 14	6
11	10	31, 20, 5, 40	5
12	15	30, 20, 23, 22, 18	4
13	14	9, 35, 16	4
14	61	15, 33, 38, 20	3
15	55	4, 41, 21	5
16	14	16, 43, 44, 31, 9	7
17	60	28, 14	2
18	9	6, 35, 30, 7, 4, 31	4
19	51	54, 40, 10	6
20	51	23, 10	5

This may be due to the underlying algorithm for the Steiner Tree problem. All the VNS algorithms converge in a superior time as compared to GA. The algorithms also produce several alternative solutions: [1 2 3 4], [1 4 3 2], [1 3 4 2], [2 1 3 4], [2 1 4 3] and [4 1 3 2].

We extend the algorithms to look at a second data set consisting of medium sized problems (scheduling 4 to 8 requests). We have selected a network consisting of 50 nodes and 83 edges that were taken from the OR-Library (steinb1 and the last 20 data from steinb2 that does not conflict with steinb1) [1]. These data sets are the benchmark problems for solving Steiner Tree problems. For simplicity, the capacity of all the edges/links was fixed at 12. The characteristics of the requests are given in Table 3.

We run all the algorithms on each data set and Table 4 tabulates the results obtained by all the algorithms. The results are based on the best of 10 runs and the time taken is an average of all the 10 runs. The algorithm also found alternative solutions for 4 requests: [3 1 4 2]; 5 requests: [5 1 3 2 4] and [5 1 2 3 4]; 6 requests: [5 1 2 3 6 4]; and 7 requests: [5 1 2 3 4 6 7]. We note

Table 6. Results for the third test case

No of Requests	Criteria	GA	VNS1	VNS2	VNS3
4R	Cost	**813**	**813**	**813**	**813**
	Time	393.5125	38.2125	30.98282	51.42346
5R	Cost	**1028**	**1028**	**1028**	**1028**
	Time	549.7438	60.74844	51.67971	117.9142
6R	Cost	**1135**	**1135**	**1135**	**1135**
	Time	599.7266	61.21408	65.7047	166.7283
7R	Cost	**1600**	**1600**	**1600**	**1600**
	Time	719.1938	78.63288	88.00941	279.6203
8R	Cost	**1768**	**1768**	**1768**	**1768**
	Time	799.7219	106.4251	87.81569	529.4485
9R	Cost	1955	1975	1969	**1951**
	Time	897.5828	102.0751	121.9547	544.1922
10R	Cost	**2272**	**2272**	2293	**2272**
	Time	1024.33	132.7064	119.2828	774.9266
11R	Cost	**2568**	2611	2646	2572
	Time	1147.95	145.2829	112.3704	708.6957
12R	Cost	2808	2897	2879	**2768**
	Time	1252.26	188.3596	118.183	1029.794
13R	Cost	2970	3058	3060	**2959**
	Time	1330.3800	158.9971	168.8579	980.7216
14R	Cost	**3190**	3252	3285	3192
	Time	1427.73	187.5609	146.1902	1312.548
15R	Cost	3412	3422	3479	**3316**
	Time	1491.8889	206.558	167.0469	1674.065
16R	Cost	**3673**	3889	4184	3900
	Time	1620.04	203.2931	219.6407	2050.567
17R	Cost	**3902**	4131	4042	4036
	Time	1700.60	252.7404	221.2611	1770.141
18R	Cost	**4053**	4479	4386	4201
	Time	1826.00	238.3081	222.8828	2442.993
19R	Cost	4974	–	–	**4601**
	Time	1906.65	–	–	2739.76
20R	Cost	**4744**	–	–	5019
	Time	1958.175	–	–	2230.875

that all the algorithms fail to converge to a feasible solution for the 8 requests problem. We observed that the capacity constraint on any of the path cannot accommodate the 8 requests. The results show that all the algorithms converge to the same objective value. The algorithms are fairly efficient in solving medium sized problems.

Encouraged by the results, we tested the algorithms on larger problems of up to 20 requests. The network comprises of 61 nodes and 133 edges. The data is shown in Table 5. The capacity for all edges is maintained at 12, and the edges and costs are taken by modifying data steinb from the OR-Library [1].

Table 6 presents the results for the larger problems and the best solutions found are highlighted in the table. We observed that the two VNS (VNS1 and VNS2) failed to converge to feasible solutions for the 19 and 20 requests problems. We note the population size for the GA algorithm has been increased to 50 individuals and the algorithms were run for 100 generations. The local search for the VNS was improved to include the *Or*-opt which have been proven to perform better for the Traveling Salesman and routing problems. This is at the expense of larger CPU time. We denote the algorithm as VNS3.

We note that GA and VNS with *Or*-opt are almost comparable with GA performing slightly better in larger problems in term of solution quality and CPU time. We note that out of the ten runs, GA produces smaller number of feasible solutions compared to VNS with *Or*-opt.

8 Conclusions

In this study, we have designed a slightly better GA as compared to GA proposed by [3]. Variable Neighborhood Search has the potential to perform better if the underlying local search is good. We observed that VNS with *Or*-opt often produces feasible solutions even for larger problems. Although GA is slightly superior in many problems as compared to VNS, it produces lower percentage of feasible solutions compared to VNS algorithms especially in larger problems. We are currently experimenting with restricted *Or*-opt in order to reduce the CPU time.

References

1. Beasley, J.E.: OR-Library: distributing test problems by electronic mail. Journal of the Operational Research Society 41, 1069–1072 (1990)
2. Blum, C., Roli, A.: Metaheuristics in Combinatorial Optimization: Overview and Conceptual Comparison. ACM Computing Surveys 35, 268–308 (2003)
3. Christensen, H.L., Wainwright, R.L., Schoenefeld, D.A.: A Hybrid Algorithm for the Point to Multipoint Routing Problem. In: Proceedings of the 1997 ACM Symposium on Applied Computing, pp. 263–268. ACM Press, New York (1997)
4. Dijkstra, E.W.: A Note on Two Problems in Connection with Graphs. Numerische Mathematik 1, 269–271 (1959)
5. Goldberg, D.E.: Genetics Algorithms in Search, Optimization, and Machine Learning. Addison Wesley Longman Inc., Canada (1999)

6. Galiasso, P., Wainwright, R.L.: A Hybrid Genetic Algorithm for the Point to Multipoint Routing Problem with Single Split paths. In: Proceedings of the 2001 ACM symposium on Applied Computing, Las Vegas, Nevada, United States, March 2001, pp. 327–332 (2001)
7. Hansen, P., Mladenović, N.: Variable Neighborhood search: Principles and applications. European Journal of Operational Research 130, 449–467 (2001)
8. Hansen, P., Mladenović, N.: Variable Neighborhood Search. In: Glover, F., Kochenberger, G.A. (eds.) Handbook of Metaheuristics, ch. 6. Kluwer Academic Publishers, Dordrecht (2003)
9. Hwang, F.K., Richards, D.S., Winter, P.: The Steiner Tree Problem. North Holland, The Netherlands (1992)
10. Kou, L., Markowsky, G., Berman, L.: A Fast Algorithm for Steiner Trees. Acta Informatica 15, 141–145 (1981)
11. Zhu, L., Wainwright, R.L., Schoenefeld, D.A.: A Genetic Algorithm for the Point to Multipoint Routing Problem with Varying Number of Requests. In: Proceeding of the 1998 IEEE International Conference on Evolutionary Computing (ICEC 1998), part of WCCI, Anchorage, Alaska (1998)

Observation-Based Interaction and Concurrent Aspect-Oriented Programming

Iulian Ober[1] and Younes Lakhrissi[1,2]

[1] Université de Toulouse - IRIT
118 Route de Narbonne, 31062 Toulouse, France
iulian.ober@irit.fr

[2] ACSYS, Faculté des Sciences de Rabat

Summary. In this paper we propose the use of *event observation* as first class concept for the composition of software components. The approach can be applied to any language based on concurrent components, and we illustrate it with examples on a concrete language (Omega UML) used in the specification of real-time systems. To motivate the proposal, we discuss how it may be used to support a very general form of aspect-oriented programming.

1 Introduction

In this paper we study the use of *event observation* as interaction mechanism between software components. The main part of the paper is concerned with the general concepts that are behind observation-based interaction. However, in the final part we are also concerned with motivating why and how this form of interaction is useful. Although other applications are possible, we particularly concentrate on how it may be used to support aspect-oriented programming.

Traditionally, event observation is employed in system modelling in very specific contexts, for example for expressing *properties* that are to be satisfied by a model or a software artifact. The origin of observers as a property specification formalism can be traced back to the Véda tool [13], and the concept has over the years proved to be both powerful and intuitive to use by non-experts in formal verification and has encountered a certain success in several other verification tools, both industrial [1] and academic [3, 4]. Our previous work [9] is a first attempt to use event observation as basis for the specification of certain aspects of a (real-time) software system, more precisely the aspects pertaining to the timing of the system execution.

In the present paper, we generalize our model from [9] in order to use event observation for behavior specification and composition (in [9] observation is only used for specifying duration constraints between events). Additionally, in this paper we propose a systematic approach for deriving event types and the definition of observations from the operational semantics of the host language to which they are added, whereas in [9] observations and event types are defined more or less ad hoc. We analyze in this paper the following questions:

R. Lee (Ed.): Soft. Eng. Research, Management & Applications, SCI 150, pp. 141–155, 2008.
springerlink.com

- How can *event observation* be smoothly integrated in a traditional computation model (the *host* model) based on communicating objects? This involves several subsidiary questions:
  What is the set of events ($\mathcal{E}$) that may be generated by the components of a system? There is no general answer to this since it depends on the specifics of the *host* model, but in §2.1 we suggest where to start searching.
  How to define the language constructs allowing a component to observe (sets of) events generated by other components? Such a construct has to capture information related to *event types* (e.g., distinguish an object creation event from a method call event), to the *scope of observation* (e.g., local to a component, global), to the existence of specific *event data* (e.g., method parameters for a method call event). The *observation* construct we define in §2.2 tries to answer these questions.
  How to compose behaviors based on observations? We tackle this question in §3.
- Once a computation model integrating observation-based interaction is defined, how can it be put to use in practice? In §4 we show that such a model can support aspect-oriented software development in a very general way.

The concrete host language that we consider in our study is a particular profile of UML [18], Omega UML [6, 9]. This choice is justified by the fact that, unlike the standard UML language, Omega UML has a well defined operational semantics given in [6] and in particular a very clear concurrency model based on a notion of active objects. Also motivating us was the fact that we previously developed execution and verification tools for the profile, which give us the possibility to rapidly prototype the concepts proposed in this paper. Nevertheless, we are concerned with the generality of our proposal and we strongly believe that the ideas presented here can be adapted to other host languages based on different computation models. Such issues are left for future exploration.

The rest of the paper begins with a short presentation of the Omega UML computation model. Afterwards, the structure of the paper is guided by the questions given previously in this section. We end it by comparing our approach to some existing proposals, and by drawing conclusions and the main lines of future work.

1.1 The Omega UML Profile

The Omega profile [6, 9] can be described as a *semantic profile* of UML, in the sense that it defines very few extensions (i.e., new language constructs), and it mainly concentrates on making precise the semantics of the existing concepts of UML. The focus is specifically on the concepts and diagrams which serve to build operational (executable) design models: class models and their associated behavior (operations and state machines). The other elements of UML (use cases, interactions, activities, deployment, etc.) are not forbidden in Omega UML models, but they are not particularly specialized, nor given a more precise semantics.

The profile specializes the semantics of UML mainly in two respects:

- The *computation model*, i.e. the runtime semantics of notions like active and passive objects (including their relationship to control threads), operation invocation, signal exchange, timing constructs, and others. Note that in this respect, the standard definition of UML intentionally leaves many questions open – these are called semantic variation points.
- The *action language* used for fine-grained description of object behavior in operations or on state machine transitions. Note that UML defines an *action semantics* with all the necessary action types, but without any concrete syntax. Therefore, any concrete application (e.g., design tool) using this part of the language has to define its own concrete language.

The choices made in Omega UML with respect to these two points are briefly outlined in this section. We emphasize that the justification of the choices made in the profile (which derive from its application domain – the design of verifiable real-time systems), is outside the scope of this paper and may be found in [6, 9]. We describe the profile only to the extent necessary to understand the examples given in the paper.

The Omega UML computation model

The OMEGA computation model is an extension of the computation model of the Rhapsody UML tool (see [10] for details). It is based on the existence of two kinds of classes: *active* and *passive* ones. At runtime, each instance of an active class defines a concurrency unit with one thread of control, called an *activity group*. Each passive class instance belongs to exactly one activity group, the one of the instance that created it.

Apart from defining the partition of the system into activity groups, there is no difference between how active and passive classes (and instances) are defined and handled. Both kinds of classes are defined by their *attributes*, *relationships*, *operations* and *state machine*, and their operational semantics is identical.

Different activity groups are considered as *concurrent*, and each activity group treats external requests (signals and operation calls from outside the group) one by one in a run-to-completion fashion, in order to protect itself from interference of concurrent requests. This protection policy is common in many concurrent computation models, including Hoare's monitors, Ada protected objects, SDL processes, etc., the particularity of Omega UML being that it is applied to a group of objects instead of only one.

As object interaction mechanism, Omega UML supports only *synchronous* operation calls, where the caller object and its activity group are blocked in a *suspended* state until the explicit return of control from the callee. Two kinds of operations are defined:

- *Triggered operations* which are not allowed to have a body and for which the behavior is described directly in the state machine of the class (the operation call is seen as a special kind of transition trigger).

- *Primitive operations* which are described by a body, like usual methods in object oriented programming languages.

The Omega action language

In order to describe a meaningful behaviour for a UML model, one also needs to describe *actions*: the effect of a transition from a state machine or the body of an operation. The OMEGA profile defines a textual action language compatible with UML action semantics [18], which covers *object creation and destruction*, *operation calls*, *expression* evaluation, *assignments*, *return* from operations, as well as control flow structuring statements (*conditionals* and *loops*). The concrete syntax is self explanatory as it relies on conventions widely used in imperative object oriented languages (Ada, C++, etc.).

An example

To illustrate the previous description we consider a small yet relatively complete example of an ATM machine in Omega UML. The structure of the model, composed of five classes, is shown in Fig. 1. The model is such that each class has

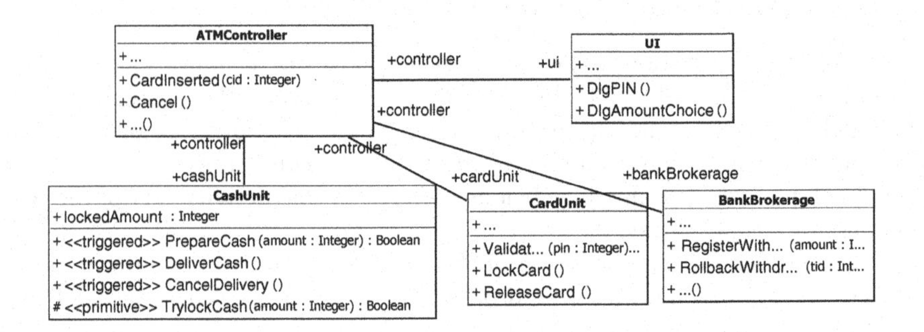

Fig. 1. ATM example: class diagram

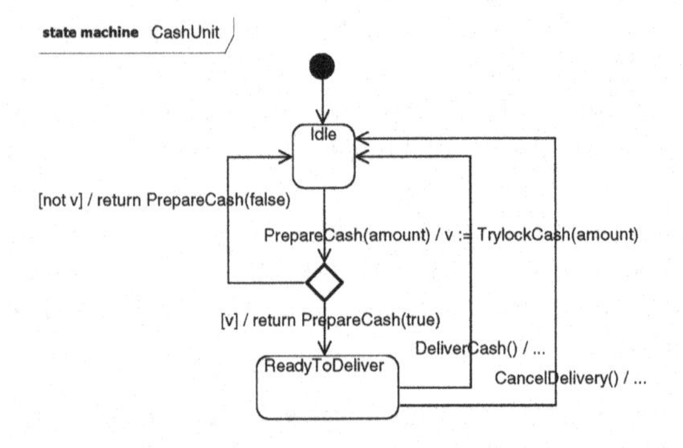

Fig. 2. ATM example: state diagram of `CashUnit`

exactly one instance at run-time; moreover, although it is not apparent in the figure, all classes are *active* – each of their instances will have its own thread of control. The structure of the classes is as usual given by attributes and relationships (associations). The behavior of each class is specified by its state machine and possibly by its *primitive* operations.

For brevity, in this example we focus in particular on one class, the `CashUnit`. Its interface is visible in Fig. 1: it contains three public triggered operations called by the `ATMController` (`PrepareCash`, `DeliverCash`, `CancelDelivery`), and one private primitive operation `TrylockCash` which groups some internal actions executed by this object (checking the cash store for availability of a requested `amount` and if ok, locking the `amount` for delivery).

The behavior of the `CashUnit` object is partially visible in Fig. 2. Initially it is in state `Idle`. Upon reception of a triggered operation call (`PrepareCash`) it calls the internal operation `TrylockCash`. Depending on the success of this operation, it may either return `false` to the caller of `PrepareCash` and go back to `Idle` state, or return `true` go to state `ReadyToDeliver`. In `ReadyToDeliver`, the object may either receive a `DeliverCash` call which causes the delivery of the locked cash (details are omitted), or it may receive a `CancelDelivery` call which causes it to restore the locked cash (this can be decided by the `ATMController` for various reasons).

2 Events and the Observation Construct

In this section we define the two essential notions of our proposal:

- *event* – a run-time entity which designates the occurrence of a particular condition in the execution of a software system/component, together with the relevant data for that occurrence.
- *observation* – a language construct which, for a given system execution, corresponds to an ordered set of *events* and allows to refer to these events in the system model (e.g., as triggers for some behavior, etc.)

2.1 Events

We relate *events* to the smallest (i.e. indivisible) state changes described by the semantics of a model/language. We consider in particular the case of languages provided with a *structural operational semantics* (SOS, in the sense of [19]), where events therefore correspond to the transitions of the semantic *labeled transition system* (LTS) associated with a model.

SOS terminology

In SOS the semantics of a program (or system model) is a labeled graph (LTS) whose vertices represent "*global states*" of the program, and whose edges (also called *transitions*) represent the smallest (atomic) steps executed by the program to go from one state to another. The graph paths which start in an identified *initial* global state represent the possible *executions* of the program.

For example, for a concurrent object-oriented language like Omega UML, a global state includes the attribute values, state and request queue for all existing objects, as well as execution context (call stack, etc.) for all threads (activity groups). An LTS transition corresponds to the execution of an individual action by one object, such as: consuming an operation request from the queue, starting to fire a state machine transition, executing an assignment, issuing an operation request, etc. Note that an LTS transition is not to be confused with a state machine transition (from the state machine associated to a UML class): the latter is usually executed as a sequence of LTS transitions.

The LTS corresponding to a program is usually not defined explicitly but rather implicitly, by a series of rules that may be used to construct it inductively. A *global state* is represented by an algebra whose signature obeys to some (naming) rules which give it a meaning. *Transitions* are not defined explicitly, but instead, a set of *transition rules* define the conditions under which a transition between two global states exists.

Omega UML events

In the case of Omega UML, the semantics involves several types of atomic steps, each defined by a specific transition rule[1]. They include: object creation, object destruction, consuming an operation call, starting to fire a state machine transition, executing an assignment, issuing an operation call, returning a result from an operation, returning control from an operation, terminating a state machine transition (i.e., entering a state), and several others.

Each LTS transition (of one of the kinds mentioned above) constitutes an event. The purpose of this work is to set out the framework allowing to use these events as means for interaction between objects.

Event data and meta-data

Every transition (event) in the SOS is defined in a particular context, and depends on a set of elements from the model (the program) and from the run-time that are specifically designated in the *transition rule* inducing it.

We take for example (Figure 3) the transition rule from the semantics of Omega UML which defines a an object performing a *triggered operation input* and subsequently firing a state machine transition. (For space reasons, we cannot include the whole SOS definition of Omega UML. The relevant elements of the rule are explained in the sequel, to the extent necessary for understanding the argument.)

[1] In order to simplify the presentation, we considered here that the semantics of Omega UML is directly defined as SOS. This is in reality not the case, the semantics being given by a set of mapping rules to a different language (IF [4]), for which, in turn, an SOS semantics is defined in [16].

$$\frac{C : q \xrightarrow{[g]op(x)/\alpha} q', \quad \begin{array}{l} \omega \in C \\ \omega.loc = q \\ \omega.v(g) = t \\ \omega.w = op(d).z \end{array}}{\Omega \xrightarrow{?op(d)} \Omega \begin{bmatrix} \omega.loc \mapsto \alpha \\ \omega.w \mapsto z \\ \omega.v \mapsto \omega.v[x \mapsto d] \end{bmatrix}}$$

Fig. 3. The *triggered operation input* transition rule

The premise of the transition rule contains the contextual elements on which the application of the rule (and hence the existence of the *event*) depends. In particular, the *triggered operation input* event described here depends on:

- The existence *in the model* of a class C, containing a state machine, containing a transition from a state q to a state q', triggered by an operation *op*, guarded by a boolean expression g and having as effect a sequence of statements α.
- The existence at run-time, *in the global system state* Ω, of an object ω of class C. ω should be precisely in state q and should have a request queue w containing (in front position) a request for *op* with a data parameter d. Moreover, the value of the guard g under the current valuation v must yield true.

Under these conditions, the transition rule states (in the bottom part) that a *triggered operation input* from the global state Ω is possible, and what is the global state after this event.

It is clear that, if the event is to be used as an interaction mechanism, i.e. it is to be *observed* by some other object, then the (run-time) *data* and the (model) *meta-data* mentioned before characterizes the event and needs also to be observable.

Note that the (meta-)data elements vary from one event type to another. For the operation input shown before, it includes: the concerned object (ω), its class (C), the machine state in which it was before the input (q), etc. For another type of event, for example *object creation*, another set of (meta-)data is relevant: the creator object, the class of the created object, the reference to the new object, etc.

We deduce that every *event* must carry this data as *parameters*, and that the observation construct (which we define below as the language mechanism for manipulating events) must offer access to the parameters. Depending on the degree of *reflectivity* available in the host language, it may or it may not be possible to talk explicitly about the meta parameters.

2.2 Observations

A model execution is a path in the LTS, and therefore corresponds to a sequence of *events* as defined before. An *observation* is a modelling construct which serves to identify and manipulate a sub-sequence of these events that is relevant for a particular goal.

Such a construct is in principle orthogonal to all the other constructs of the *host* language (Omega UML in our case) – in the sense that it serves a different purpose and it is not structurally related (contained into, etc.) to any other existing language construct. Later we will see that it is possible to relate observations to existing constructs (e.g., for specifying a context in which observed events should occur), but this is not in any way fundamental to the notion of observation and is seen as syntactic sugar.

Observations are used to achieve observation-based interaction between objects (see below in §3), but in order to do so they also have to provide access to the event (meta-)data as outlined in §2.1. This is achieved by considering observations as being objects themselves, with attributes that store the relevant data of the last event matching them at any time during execution. A very good term of comparison from commonly used languages are *exception objects* in languages like C++ or Java: they can be used for communication and for triggering behavior (when they are caught), but they are also objects themselves, with attributes, etc. To continue the comparison, observations can be seen as a sort of exceptions raised by default (when a matching event occurs), which can be "caught" and thus trigger some behavior, but which do not have the disruptive quality of "usual" exceptions if they are not caught.

We argue that all *observations* that are interesting for modelling can be constructed based on a limited number of *elementary observations* (roughly corresponding to the transition rules of the SOS) and on a set of *operations on observations* which are derived from the most common set operations (union, intersection, complementing, projection).

Elementary observations

We stated before that events correspond to transitions from the semantic LTS. As such, each event has a *kind*, which is the *transition rule* of the SOS from which it is derived. It follows logically that we can define elementary *observations* which correspond to these event *kinds*. If the sequence of events generated by an execution is $\mathcal{E} = (e_1, e_2, ...)$, then an *elementary observation* $\mathcal{O}$ generates a sub-sequence $\mathcal{E}_\mathcal{O} = (e_{k_1}, e_{k_2}, ...)$ such that all e_{k_i} are of kind $\mathcal{O}$.

In the case of Omega UML, for example, to the *triggered operation input* rule in Figure 3 corresponds an elementary observation $\mathcal{O}_{toinput}$. Note that the number of rules is usually quite small even for complex languages (around 15 for Omega UML depending on how the semantics is defined), and that all types of events are not necessarily interesting to observe.

Observation operators

Remember that the semantic model of an observation is a subset of the whole sequence of events generated in a system execution ($\mathcal{E}$). It is then natural to allow observations to be composed using set-based operators: union, intersection, complementation, projection.

The semantics of observation union, intersection and complementation is self-explanatory. The projection operation allows to obtain from an observation $\mathcal{O}$ (with semantic model $\mathcal{E}_{\mathcal{O}}$) another observation $\mathcal{O}'$ whose semantic model $\mathcal{E}_{\mathcal{O}'}$ is a subsequence of $\mathcal{E}_{\mathcal{O}}$, based on a boolean condition on the event parameters.

Here are some examples of use of the operators (for simplicity, we use the standard mathematical notations for observation operators):

- for identifying the triggered operation inputs that affect only a particular variable x one can take a projection of the elementary observation $\mathcal{O}_{toinput}$ (suppose the *.store* meta-data is a list of the attributes affected by an input):

$$\mathcal{O}_{toinput(x)} = \mathcal{O}_{toinput}\big|_{x \in .store}$$

- for identifying the events that lead to modifying a particular variable x, one can take the union of $\mathcal{O}_{toinput(x)}$ (defined before) and of the observation of *assignments* which affect x:

$$\mathcal{O}_{mod\,x} = \mathcal{O}_{toinput(x)} \quad \cup \quad \mathcal{O}_{assign}\big|_{x \in .store}$$

Implicit projections

Often one is interested in observing a particular kind of events only in a particular context (e.g., the *inputs* performed by one particular object). In principle, the projection operator presented before is sufficient for specifying the context, although in practice this may be syntactically awkward (for our example, one needs to explicitly designate the identity of the concerned object, raising the question of how this identity is obtained).

Therefore, it may sometimes be useful to define observations in the program not as top-level constructs but within a context (e.g., inside a class), meaning that there is an implicit projection of the observation based on the context. This does not augment the expressive power of observations defined before, and is only syntactic sugar.

3 Behavior Composition Using Observations

The purpose of introducing *observations* is to be able to identify *events* and use them in the behavioral specification of components (objects in Omega UML). The way to do this depends on the computational model assumed by the host language. In the following, we propose a solution which works for Omega UML and can presumably be adapted to any concurrent language (synchronous or asynchronous).

The elementary extension that is necessary is a statement which, for a given observation o, waits until an event matching o is produced. Let $wait(o)$ denote this statement. Since this statement is possibly blocking, it has to be introduced in a way which is consistent to the language definition: for example in Omega UML the assumption is that the only blocking statements are the *transition triggers*. Therefore, $wait(o)$ has to be defined as a new type of transition trigger.

Semantics of observation waiting

Clarifying the semantics of the $wait(o)$ trigger mainly means answering the following question: after an event e that matches o has occurred, when (and for how long) is $wait(o)$ enabled (i.e., able to fire the associated transition).

When a component (object) c_1 executes an action which induces an event e matching an observation o which is waited upon by a second component c_2 (see Figure 4) this implies an *implicit communication* taking place between c_1 and c_2. Prior to this communication, the attributes of o are updated (with the event data). It is a semantic choice whether to consider this communication *synchronous* or *asynchronous*. In the former case, the *wait* transition can be fired only if c_2 is already in state q_2 when e occurs. c_2 becomes executable immediately (in terms of threads, it goes from suspended to executable). In the latter, the *wait* transition may be fired at a later moment, including after c_2 has executed some other actions, and it implies that the run-time has the ability to keep the memory of the event for every component until it able to "consume" (observe) it.

Unlike for normal message-based communication, where asynchronous and synchronous messages have the same expressive power (i.e. each one can be expressed with the other, see for ex. [5]), for observations the synchronous semantics is *strictly more powerful* in the following sense: given an asynchronous observation mechanism, it is impossible to express a synchronous observation without modifying the specification of the event *source* component (c_1 above). (An explicit waiting for an acknowledgement has to be inserted in c_1 in order to achieve synchrony.) Since the reason of existence of the observation-based interaction is precisely to make possible for c_2 to react to events *implicitly* produced by c_1, requiring modifications in c_1 in order for it to work would be self-defeating.

In consequence, we settle for the choice of *synchronous* observations. For space reasons, we refrain from giving the SOS rules defining event production and observation in Omega UML. However, the precise semantics can be understood by comparison to known synchronization constructs from other frameworks, in particular to *condition variables* from Hoare's monitors [11] (also known from the Java language via the `wait` and `notifyAll` functions of `java.lang.Object`). The semantics of *wait(o)* is identical to waiting on a *condition variable* associated to o. Event generation is the (atomic execution of) updating o's attributes with the actual event parameters, followed by signaling the condition variables associated to all matching observations (o,o',o'' in Fig. 4).

We note that based on the synchronous semantics it is possible to achieve stronger constraints if necessary: for example if, upon e, c_2 has to take a series

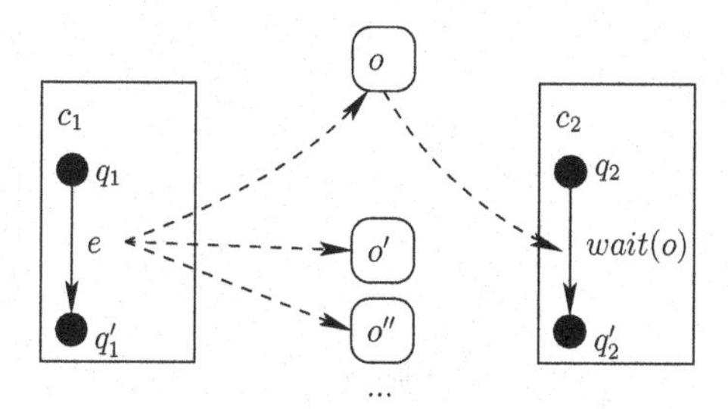

Fig. 4. Observation-based communication

of actions *before* c_1 *may continue*, this can be achieved simply by giving c_2 a higher priority compared to c_1. To do so, one may use either the capabilities of the run-time (priority levels are available for example in all real-time operating systems), or constructs available in the host language (e.g., *dynamic priority rules* as they exist in Omega UML [9]). Although this kind of combination of language features is important, we cannot focus more on it in this paper.

4 Observers as Dynamic Aspects

Aspect orientation [15] is based on the idea that, in the process of designing a (software) system, certain concerns or requirements may impact several components of the established system architecture. Such concerns are called *cross-cutting aspects.* Aspect-oriented languages usually consist of a traditional (object oriented, procedural or functional) host language for specifying the architecture and the basic system functionality, and of a set of constructs for specifying aspects.

An aspect is generally defined in terms of the aspect's *joinpoints* – the program elements that are affected by the aspect and the conditions under which this happens, and the aspect's *advice* – the structural and/or behavioral elements added by the aspect. The semantics of an aspect-oriented language is given by the rules for composing (*weaving*) aspects and the basic system specification into an executable model.

The *observation* constructs defined in the previous sections may be used as a basis for defining joinpoints: the execution of aspect-related code is then triggered by the occurrence of events. By combining this with the control-flow constructs already available in a host language such as Omega UML, aspects can be triggered not only by individual events but by arbitrary patterns of events, with arbitrary conditions based on event data, etc. This is in general more expressive than what can be achieved in most "classical" aspect-oriented models, like the one of AspectJ [14], in which *joinpoints* are based mostly on syntactic conditions.

We illustrate the approach with a commonly used example, which adds a *logging* functionality to an existing system – in our case the ATM described in §1.1. Our logging differs from more classical examples in that it must take place

$\mathcal{O}_{ctl} = \mathcal{O}_{pocall} \big|_{.op = \texttt{CashUnit::TrylockCash} \wedge .pars[0] \geq 100}$

$\mathcal{O}_{rtltrue} = \mathcal{O}_{poreturn} \big|_{.op = \texttt{CashUnit::TrylockCash} \wedge .ret = \texttt{true}}$

$\mathcal{O}_{rtlfalse} = \mathcal{O}_{poreturn} \big|_{.op = \texttt{CashUnit::TrylockCash} \wedge .ret = \texttt{false}}$

$\mathcal{O}_{deliver} = \mathcal{O}_{toinput} \big|_{.op=\texttt{CashUnit::DeliverCash}}$

$\mathcal{O}_{cancel} = \mathcal{O}_{toinput} \big|_{.op=\texttt{CashUnit::CancelDelivery}}$

$\mathcal{O}_{dc} = \mathcal{O}_{deliver} \cup \mathcal{O}_{cancel}$

state machine logger

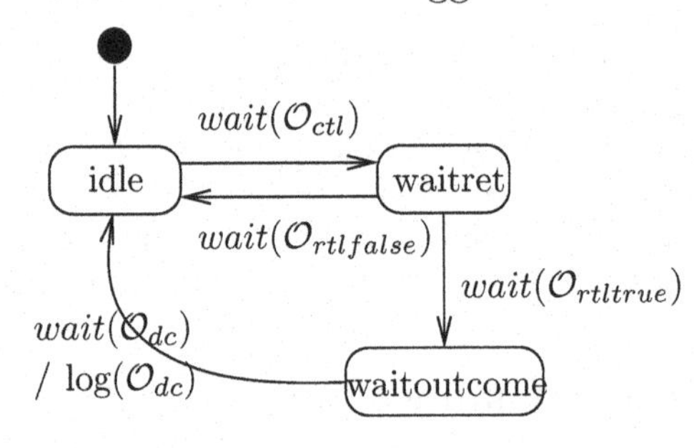

Fig. 5. A logger aspect based on observations

under some run-time conditions which involve the occurrence of certain events in a specific order. While remaining fairly simple, this example suggests what can be achieved by combining observations with full fledged state machines.

We consider the following requirement: each time the cash unit successfully locks an amount of cash greater than 100 for delivery, the ATM must subsequently write a log entry with the outcome of the transaction (whether it was canceled or whether the cash was effectively delivered). Fig. 5 shows the specification of this aspect; for simplicity, we use the mathematical notation introduced in §2.2, the actual model-based notation differing from it only syntactically.

The aspect is implemented as a separate active object, with its own state machine. It uses the following observations: $\mathcal{O}_{ctl}$ matching the calls to `TrylockCash` with an amount higher than 100, constructed from the elementary observation $\mathcal{O}_{pocall}$ which denotes all calls to primitive operations. $\mathcal{O}_{rtltrue}$ and $\mathcal{O}_{rtlfalse}$ matching the return from `TrylockCash` successfully, respectively unsuccessfully (constructed from the elementary observation $\mathcal{O}_{poreturn}$ which denotes all primitive operation returns). $\mathcal{O}_{deliver}$ and $\mathcal{O}_{cancel}$ denote the triggered operation inputs for `DeliverCash` and respectively `CancelDelivery`, and $\mathcal{O}_{dc}$ is the union of the two.

Based on these, the control structure of the aspect is as follows: whenever it observes a call to `TrylockCash` with an amount higher than 100 and which returns `true`, it waits for an input of either `DeliverCash` or `CancelDelivery`

($\mathcal{O}_{dc}$) and when this occurs it logs the related information (which can be retrieved directly from $\mathcal{O}_{dc}$).

We have applied this form of aspects to model other cross-cutting concerns, such as *alarm management* in an embedded controller. Alarm management involves dynamic conditions on the order of events, their data and also their timing (e.g., an alarm has to be issued if the plant does not reach some operating state within n milliseconds from a *start* event). The construction is similar (if more complex) to the one in the logger example.

Finally, we note that with this approach, nothing distinguishes an aspect from a "normal" system component, and one can mix observation-based behavior and conventional communication-based behavior inside the same component. This opens the way to a methodology of system specification in which well-behaved (decomposable) concerns are implemented across the component architecture by conventional means, while ill-behaved (cross-cutting) concerns are implemented using observations. Such a methodology would approach the goals put forward by Jacobson and Ng [12] in their *use-case slicing* method, which has stimulated much interest in recent years, but for which current aspect specification frameworks are generally considered not powerful enough.

5 Related Work and Conclusions

The concepts of event and observation presented here are an evolution of those we described in [9]. There they served as basis for defining a particular type of aspects (execution timing constraints) by using a dedicated *declarative* language, which is less flexible than the generalized use of observations as interaction mechanism, proposed here.

A criticism against observation-based interaction is that it appears to lead to tightly coupled component architectures. The fact is that implementing cross-cutting concerns inherently leads to tightly coupled components, and aspect-oriented languages aim at reducing the burden of the designer by making part of the coupling *implicit*. Our approach does the same: when the coupling between two components is unidirectional (observed-observer), it provides a way to avoid the impact on the specification of the observed component.

The body of literature dedicated to aspect models and languages is wide and growing rapidly. An important classification criterion for aspect frameworks concerns the constructs for defining *join points* and their expressive power. From this point of view, most of the proposed frameworks (including that of AspectJ [14]) use syntactic joinpoint models. However, a few recent proposals are closer to ours, in that they propose event (trace) based joinpoint models [20, 2, 8], sometimes combined with a form of dynamic (run-time) weaving (e.g., [7, 17]).

In [20, 2, 17], join points (sometimes called *tracecuts* because they are based on a multi-event trace) are defined in a declarative way, for example as a regular expression on events, and concern only sequential programs. The computational model that is closest to ours is the Concurrent Event-Based AOP (CEAOP) defined in [7], and which is based on the same principles of parallel composition of system components and aspects, and of event based synchronization. The main difference is that we define the characteristics of events and the observation construct (§2), whereas in [7] events are just simple synchronization labels. Also, in [7] the authors distinguish normal system components and aspects (although their computation model, like ours, does not require this distinction). We think that not making the distinction, and uniformly offering the same observation mechanisms to all components can lead to interesting results in keeping concerns related to each use-case separate inside the component descriptions (this technique is known as *use case slicing* [12]). Further evidence is needed to support this claim.

Still concerning future work, we note that the observation concept (in particular, access to event meta-data) requires reflective support in the language in order to be fully satisfactory. Therefore, we plan to test the concepts presented here on a language which offers such support, together with support for asynchronous concurrency, such as some extensions of Python.

The observation mechanisms we originally proposed in [9] had been implemented in a centralized way, in the execution platform. This is possible also in the present setting and it is acceptable since the Omega profile is mostly dedicated to simulation and formal property verification, where a centralized event monitor does not affect performance. If the target is final implementation, event observation must be done more efficiently, e.g. by code instrumentation based on a publish-subscribe mechanism. The work of [7], on a similar model, shows that this is in principle possible.

References

1. Algayres, B., Lejeune, Y., Hugonnet, F.: GOAL: Observing SDL behaviors with ObjectGEODE. In: Braek, R., Sarma, A. (eds.) SDL 1995: with MSC in CASE, Proceedings of the 7th SDL Forum. Elsevier Science, B.V. (1995)
2. Allan, C., Avgustinov, P., Christensen, A.S., Hendren, L.J., Kuzins, S., Lhoták, O., de Moor, O., Sereni, D., Sittampalam, G., Tibble, J.: Adding trace matching with free variables to aspectj. In: Johnson, R., Gabriel, R.P. (eds.) OOPSLA, pp. 345–364. ACM, New York (2005)
3. Bozga, M., Graf, S., Mounier, L.: If-2.0: A validation environment for component-based real-time systems. In: Brinksma, E., Larsen, K.G. (eds.) CAV 2002. LNCS, vol. 2404, pp. 343–348. Springer, Heidelberg (2002)
4. Bozga, M., Graf, S., Ober, I., Ober, I., Sifakis, J.: The if toolset. In: Bernardo, M., Corradini, F. (eds.) SFM-RT 2004. LNCS, vol. 3185, pp. 237–267. Springer, Heidelberg (2004)
5. Burns, A., Wellings, A.: Real-Time Systems and Programming Languages, 3rd edn. Addison-Wesley, Reading (2001)
6. Damm, W., Josko, B., Pnueli, A., Votintseva, A.: Understanding uml: A formal semantics of concurrency and communication in real-time UML. In: de Boer, F.S., Bonsangue, M.M., Graf, S., de Roever, W.-P. (eds.) FMCO 2002. LNCS, vol. 2852, pp. 71–98. Springer, Heidelberg (2003)
7. Douence, R., Le Botlan, D., Noyé, J., Südholt, M.: Concurrent aspects. In: Jarzabek, S., Schmidt, D.C., Veldhuizen, T.L. (eds.) GPCE, pp. 79–88. ACM, New York (2006)

8. Filman, R.E., Havelund, K.: Realizing aspects by transforming for events. Technical Report 02.05, RIACS (September 2002); Presented at ASE 2002 Workshop on Declarative Meta-Programming
9. Graf, S., Ober, I., Ober, I.: A real-time profile for UML. STTT 8(2), 113–127 (2006)
10. Harel, D., Kugler, H.: The rhapsody semantics of statecharts (or, on the executable core of the UML) - preliminary version. In: Ehrig, H., Damm, W., Desel, J., Große-Rhode, M., Reif, W., Schnieder, E., Westkämper, E. (eds.) INT 2004. LNCS, vol. 3147, pp. 325–354. Springer, Heidelberg (2004)
11. Hoare, C.A.R.: Monitors: An operating system structuring concept. Commun. ACM 17(10), 549–557 (1974)
12. Jacobson, I., Ng, P.-W.: Aspect-Oriented Software Development with Use Cases. Object Technology Series. Addison-Wesley, Reading (2005)
13. Jard, C., Monin, J.-F., Groz, R.: Development of Véda, a prototyping tool for distributed algorithms. IEEE Trans. Software Eng. 14(3), 339–352 (1988)
14. Kiczales, G., Hilsdale, E., Hugunin, J., Kersten, M., Palm, J., Griswold, W.G.: An overview of aspectj. In: Knudsen, J.L. (ed.) ECOOP 2001. LNCS, vol. 2072, pp. 327–353. Springer, Heidelberg (2001)
15. Kiczales, G., Lamping, J., Mendhekar, A., Maeda, C., Lopes, C.V., Loingtier, J.-M., Irwin, J.: Aspect-oriented programming. In: Aksit, M., Matsuoka, S. (eds.) ECOOP 1997. LNCS, vol. 1241, pp. 220–242. Springer, Heidelberg (1997)
16. Marius Bozga, Y.L.: IF-2.0 common language operational semantics. Technical report, VERIMAG (September 2002), `http://www-if.imag.fr`
17. Benavides Navarro, L.D., Südholt, M., Vanderperren, W., De Fraine, B., Suvée, D.: Explicitly distributed AOP using AWED. In: Filman, R.E. (ed.) AOSD, pp. 51–62. ACM, New York (2006)
18. Object Management Group. Unified Modeling Language, `http://www.omg.org/spec/UML/`
19. Plotkin, G.D.: A structural approach to operational semantics. J. Log. Algebr. Program. 60(61), 117–139 (2004)
20. Walker, R.J., Viggers, K.: Implementing protocols via declarative event patterns. In: Taylor, R.N., Dwyer, M.B. (eds.) SIGSOFT FSE, pp. 159–169. ACM, New York (2004)

Modular Compilation of a Synchronous Language

Annie Ressouche[1], Daniel Gaffé[2], and Valérie Roy[3]

[1] INRIA Sophia Antipolis - Méditerranée
2004 route des Lucioles BP 93 06902 Sophia Antipolis, France
Annie.Ressouche@sophia.inria.fr

[2] LEAT Laboratory, Univ of Nice Sophia Antipolis CNRS
250 rue Albert Einstein 06560 Valbonne France
daniel.gaffe@unice.fr

[3] CMA Ecole des Mines Sophia Antipolis France
vr@cma.ensmp.fr

Summary. Synchronous languages rely on formal methods to ease the development of applications in an efficient and reusable way. Formal methods have been advocated as a means of increasing the reliability of systems, especially those which are safety or business critical. It is still difficult to develop automatic specification and verification tools due to limitations like state explosion, undecidability, etc... In this work, we design a new specification model based on a reactive synchronous approach. Then, we benefit from a formal framework well suited to perform compilation and formal validation of systems. In practice, we design and implement a special purpose language (LE) and its two semantics : the *behavioral semantics* helps us to define a program by the set of its behaviors and avoid ambiguousness in programs' interpretation; the *execution equational semantics* allows the modular compilation of programs into software and hardware targets (C code, Vhdl code, Fpga synthesis, Verification tools). Our approach is pertinent considering the two main requirements of critical realistic applications : the modular compilation allows us to deal with large systems, the model-driven approach provides us with formal validation.

1 Introduction

Synchronous languages [3, 1, 8] have been designed to specify reactive systems [10]. All are model-driven languages to allow both efficiency and reusability of system design, and formal verification of system behavior. They rely on the *synchronous hypothesis* which assumes a discrete logic time scale, made of instants corresponding to reactions of the system. All the events concerned by a reaction are simultaneous : input events as well as the triggered output events. As a consequence, a reaction is instantaneous (we consider that a reaction takes no time, in compliance with synchronous language class), there are no concurrent partial reactions and so determinism can be ensured.

Although synchronous languages have begun to face the state explosion problem, there is still a need for further research on efficient and modular compilation of synchronous languages. The first compilers translated the program into an

R. Lee (Ed.): Soft. Eng. Research, Management & Applications, SCI 150, pp. 157–171, 2008.
springerlink.com

extended finite state machine. The drawback of this approach remains the potential state explosion problem. Polynomial compilation was first achieved by a translation to equation systems that symbolically encode the automata. This approach is the core of commercial tool [17]. Then several approaches translate the program into event graphs [19] or concurrent data flow graphs [6, 14] to generate efficient C code. All these methods have been used to optimize the compilation times as well as the size and the execution of the generated code.

However none of these approaches consider a modular compilation. Of course there is a fundamental contradiction in relying on a formal semantics to compile reactive systems because a perfect semantics would combine three important properties: *responsiveness*, *modularity* and *causality*. Responsiveness means that we can deal with a logical time and we can consider that output events occur in the same reaction that the input events causing them. It is one of the foundations of the synchronous hypothesis. Causality means that for each event generated in a reaction, there is a causal chain of events leading to this generation. No causal loop may occur. A semantics is modular when "environment to component" and "component to component" communication is treated symmetrically [11]. In particular, the semantics of the composition of two reactive systems can be deduced from the respective semantics of each sub-part. Another aspect of modularity is the coherent view each subsystem has of what is going on. When an event is present, it is broadcasted all around the system and is immediately available for every part which listen to it. Unfortunately, there exists a theorem ("the RMC barrier theorem") [11] that states that these three properties cannot be united in a semantics. Synchronous semantics are responsive and modular. But causality remains a problem in these semantics and modular compilation must be completed by a global causality checking.

In this paper we introduce a reactive synchronous language, we define its behavioral semantics that gives a meaning to programs and an equational semantics allowing first a modular compilation of programs and second an automatic verification of properties. As other synchronous semantics, we get a causality problem and we face it with the introduction of a new sorting algorithm that allows us to start from compiled subsystems to compile the overall system without sort again all the equations. The paper is organized as follows: section 2 is dedicated to LE language: Its syntax is briefly described and its behavioral and equational semantics are both discussed. Section 3 details how we perform a separated compilation of LE programs. Then, we compare our approach with others in section 4. Finally, we conclude and open up the way for future works in section 5.

2 LE Language

2.1 Language Overview

LE language belongs to the family of reactive synchronous languages. It is a discrete control dominated language. Nevertheless, we benefit from a great many studies about synchronous language domain since two decades. As a consequence,

Table 1. LE Operators

nothing	does nothing
emit speed	signal *speed* is immediately present in the environment
present S { P1} else { P2}	If signal S is present $P1$ is performed otherwise $P2$
$P_1 \gg P_2$	perform P_1 then P_2
$P_1 \| P_2$	synchronous parallel: start P_1 and P_2 simultaneously and stop when both have terminated
abort P when S	perform P until an instant in which S is present
loop $\{P\}$	perform P and restart when it terminates
local S $\{P\}$	encapsulation, the scope of S is restricted to P
Run M	call of module M
pause	stop until the next reaction
waitS	stop until the next reaction in which S is present
$\mathcal{A}(\mathcal{M}, \mathcal{T}, \mathcal{C}ond, M_f, \mathcal{O}, \lambda)$	automata specification

we choose to not introduce the powerful trap exit mechanism existing in the Esterel synchronous language [3] since it is responsible for a large part of the complexity of compilation. We just keep an abortion operator that only allows to exit one block. It is not a strong restriction since we can mimic trap exit with cascade of abortions. On the other hand, our language offers an *automaton description* as a native construction. Moreover, our graphical tool (GALAXY) helps the user editing automata and generating the LE code.

More precisely, LE language unit is a *named module*. The *module interface* declares the set of *input events* it reacts to and the set of *output events* it emits. In addition, the location of external already compiled sub modules is also specified in module interface. The *module body* is expressed using a set of *operators*. The language's operators and constructions are chosen to fit the description of reactive applications as a set of concurrent communicating subsystems. Communication takes place between modules or between a module and its environment. Subsystems communicate via *events*. Besides, some operators (wait, pause) are specially devoted to deal with the logical time. We do not detail LE operators, we define them in table 1 and a complete description can be found in [15]. But, we just underline the presence of the *run module* operator that calls an external module and supports a renaming of the interface signals of the called module. It is through this operator usage that modular compilation is performed.

2.2 LE Semantics

Now, we discuss LE semantics. Our approach is twofold: first, we define a *behavioral* semantics providing a consistent meaning to each LE program. Such a semantics defines a program by the set of its behaviors. Second, we propose an *equational* semantics to express programs as a set of equations and get a compilation means. We introduce these two semantics because we need both a well suited framework to apply verification techniques and an operational framework to get an effective compilation of programs.

Mathematical Context

Similarly to others synchronous reactive languages, LE handles *broadcasted signals* as communicating means. A program reacts to input events in producing output events. An *event* is a signal carrying some information related to its *status*. To get an easier way to check causality, we introduce a boolean algebra (called ξ) that provides us with a smarter information about event. ($\xi = \{\perp, 0, 1, \top\}$). Let S be a signal, S^x denotes its status in a reaction. More precisely, S^1 means that S is present, S^0 means that S is absent, $S^\perp$ means that S status has not been established (neither from the external context nor from internal information propagation), and finally $S^\top$ corresponds to an event the status of which cannot be induced because it has two incompatible status in two different sub parts of the program. For instance, if S is both absent and present, then it turns out to have $\top$ status and thus an error occurs. Indeed the set ξ is a complete lattice[1].

We define three internal composition laws in ξ: $\sqcup$, $\sqcap$and $\neg$. The $\sqcup$ law yields the upper bound of its two operands. The $\sqcap$law yields the lower bound of its two operands, while the $\neg$ law is an inverse law. The set ξ with these 3 operations verify the axioms of *Boolean Algebra.* As a consequence, we can apply classical results concerning Boolean algebras to solve equation systems whose variables belong to ξ. The equational semantics described in section 2.2 relies on boolean algebra properties to compute signal status as solution of ξ equations.

An *environment* E is a set of events built from an enumerable set of signals, where each signal has a single status (i.e if S^x and $S^y \in E$ then $x = y$). Environments are useful to record the current status of signals in a program reaction. We easily extend the operation defined in ξ to environments[2]. We define relation ($\preceq$) on environments as follows:

$$E \preceq E' \text{ iff } \forall S^x \in E, \exists S^y \in E' | S^x \leq S^y$$

Thus $E \preceq E'$ means that each element of E is less than an element of E' according to the lattice order of ξ. As a consequence,the set of environments built from a signal set S ordered with the $\preceq$ relation is a lattice and $\sqcup$ and $\sqcap$operations are monotonic with respect to $\preceq$.

[1] With respect to the order($\leq$): $\perp \leq 0 \leq \top$; $\perp \leq 1 \leq \top$; $\perp \leq \top$.

[2] Let E and E' be 2 environments:

$$\begin{aligned}
E \sqcup E' &= \{S^z | \exists S^x \in E, S^y \in E', z = x \sqcup y\} \\
&\cup \{S^z | S^z \in E, \nexists S^y \in E'\} \\
&\cup \{S^z | S^z \in E', \nexists S^y \in E\} \\
E \sqcap E' &= \{S^z | \exists S^x \in E, S^y \in E', z = x \sqcap y\} \\
\neg E &= \{S^x | \exists S^{\neg x} \in E\}
\end{aligned}$$

LE Behavioral Semantics

To define the behavioral semantics of LE programs, we formalize the notion of concurrent computation and we rely on an algebraic theory that allows the description of behaviors in a formal way. The behavioral semantics formalizes a reaction of a program P according to an event input set. $(P, E) \longmapsto (P', E')$ has the usual meaning: E and E' are respectively input and output environments; program P reacts to E, reaches a new state represented by P' and the output environment is E'. This semantics supports a rule-based specification to describe the behavior of LE statement. A rule of the semantics has the form: $p \xrightarrow[E]{E', TERM} p'$ where p and p' are LE statements. E is an environment that specifies the status of the signals declared in the scope of p, E' is the output environment and $TERM$ is a boolean termination flag. Let P be a LE program and E an input event set, a reaction is computed as follows:

$$(P, E) \longmapsto (P', E') \quad \text{iff} \quad \Gamma(P) \xrightarrow[E]{E',\ TERM} \Gamma(P')$$

where $\Gamma(P)$ is the LE statement body of program P. We cannot detail all semantics definition due to a lack of space. To give a flavor of semantics rules, we show the rule for parallel operator. This operator respects the synchronous paradigm. It computes its two operands according to the broadcast of signals between each side and it is terminated when both sides are.

$$\frac{p \xrightarrow[E]{E_p,\ TERM_p} p' \quad , \quad q \xrightarrow[E]{E_q,\ TERM_q} q'}{p \| q \xrightarrow[E]{E_p \sqcup E_q,\ TERM_p . TERM_q} p' \| q'} \quad (parallel)$$

The behavioral semantics is a "macro" semantics that gives the meaning of a reaction for each LE statement. Nevertheless, a reaction is the least fixed point of a micro step semantics that computes the output environment from the initial one. At each micro step, the environment is increased using the $\sqcup$ operation. According to the monotonicity of $\sqcup$ with respect to the $\preceq$ order and to the lattice character of the environment sets based on a given signal set, we can ensure that for each term, this least fixed point exists.

LE Equational Semantics

The behavioral semantics describes how the program reacts in an instant. It is logically correct in the sense that it computes a single output environment for each input event environment when there is no causality cycles. To face this inherent causality cycle problem specific to synchronous approach, constructive semantics have been introduced [2]. Such a semantics for synchronous languages is the application of constructive boolean logic theory to synchronous language semantics definition. The idea of constructive semantics is to "forbid self-justification and any kind of speculative reasoning replacing them by a fact-to-fact propagation" [2]. A program is *constructive* if and only if fact propagation is sufficient to establish the presence or absence of all signals. An elegant means to define a constructive semantics for a language is to translate each program into a constructive circuit. Such a translation ensures that programs containing no cyclic instantaneous signal dependencies are translated into cycle free circuits.

LE equational semantics respects this constructive principle. It computes output environments from input ones according to the ξ sequential circuits we associate with LE programs. Environments are built from the disjoint union of (1) a set of input, output and local signals, (2) a set of wires $\mathcal{W}$ and (3) a set of registers R (initially valued to 0). Registers are memories that feed back the circuit. The translation of a program into a circuit is structurally done. Sub-circuits are associated with program sub-terms. To this aim, we associate a circuit with each LE statement. Given p a LE statement, we call $\mathcal{C}(p)$ its associated ξ circuit. Each circuit $\mathcal{C}(p)$ has three interface wires: Set_p input wire activates the circuit, Reset_p input wire deactivates it and RTL_p is the "ready to leave" output wire. This latter indicates that the statement can terminate in the current clock instant. Reset_p wire is useful to deactivate a sub-circuit and top down propagate this deactivation. For instance, consider the sequence operator $(P_1 \gg P_2)$. Its circuit $\mathcal{C}(P_1 \gg P_2)$ is composed of $\mathcal{C}(P_1)$, $\mathcal{C}(P_2)$ and some equations defining its interface wires. The $\mathrm{Set}_{P_1 \gg P_2}$ input wire forwards down the control (Set_{P_1} $= \mathrm{Set}_{P_1 \gg P_2}$). The $\mathrm{Reset}_{P_1 \gg P_2}$ input wire on one hand deactivates $\mathcal{C}(P_1)$ when its computation is over, and on the other hand, forwards down the deactivation information ($\mathrm{Reset}_{P_1} = \mathrm{Reset}_{P_1 \gg P_2} \sqcup \mathrm{RTL}_{P_1}$). Finally, the overall circuit is ready to leave when P_2 is ($\mathrm{RTL}_{P_1 \gg P_2} = \mathrm{RTL}_{P_2}$). Moreover, a circuit can need a register (denoted R) when the value of a wire or a signal is required for the next step computation.

The computation of circuit output environments is done according to a propagation law and to ensure that this propagation leads to logically correct solutions, a constructive value propagation law is supported by the computation and solutions of equation systems allow us to determine all signal status.

Let $\mathcal{C}$ be a circuit, E an input environment, the constructive propagation law ($\hookrightarrow$) has the form : $E \vdash w \hookrightarrow bb$, where w is a ξ expression and bb is a ξ value; the law means that from E assignment values to signals and registers, w evaluates to bb. The $\hookrightarrow$ law is defined in table 2. By extension, let $\mathcal{C}$ be a circuit, $E \vdash \mathcal{C} \hookrightarrow E'$ means that E' is the result of the application of the propagation law to each equations of the circuit.

Now, we define the equational semantics $\mathcal{S}_e$. It is a mapping that computes an output environment from an input one and a LE statement. Let p be a LE statement, and E an environment, $\mathcal{S}_e(p, E) = \langle p \rangle_E$ iff $E \vdash \mathcal{C}(p) \hookrightarrow \langle p \rangle_E$. Finally, Let P be a LE program, the equational semantics also formalizes a reaction of a program P according to an input environment:$(P, E) \longmapsto E'$ iff $\mathcal{S}_e(\Gamma(P),$ $E) = E'$.

To illustrate how the output environment is built and similarly to section 2.2, we focus on the parallel LE operator. As already said, each circuit $\mathcal{C}(p)$ has three specific connexion wires belonging to $\mathcal{W}$ (Set_p, Reset_p and RTL_p). For the parallel operator $(P_1 \| P_2)$, the output environment $\langle P_1 \| P_2 \rangle_E$ is the upper bound

Table 2. Definition of the constructive propagation law ($\hookrightarrow$). The value between square brackets ([x]) means "respectively". For instance: $E \vdash e \hookrightarrow 1[0]$ and $E \vdash e' \hookrightarrow 0[1]$ means $E \vdash e \hookrightarrow 1$ and $E \vdash e' \hookrightarrow 0$ or $E \vdash e \hookrightarrow 0$ and $E \vdash e' \hookrightarrow 1$. $E(w)$ denotes the value of w in E.

$$E \vdash bb \hookrightarrow bb \qquad \frac{E(w) = bb}{E \vdash w \hookrightarrow bb} \qquad \frac{E \vdash e \hookrightarrow bb}{E \vdash (w = e) \hookrightarrow bb} \qquad \frac{E \vdash e \hookrightarrow \neg bb}{E \vdash \neg e \hookrightarrow bb}$$

$$\frac{E \vdash e \hookrightarrow \top \text{ or } E \vdash e' \hookrightarrow \top}{E \vdash e \sqcup e' \hookrightarrow \top} \qquad \frac{E \vdash e \hookrightarrow \bot \text{ or } E \vdash e' \hookrightarrow \bot}{E \vdash e \sqcap e' \hookrightarrow \bot}$$

$$\frac{E \vdash e \hookrightarrow 1[0] \text{ and } E \vdash e' \hookrightarrow 0[1]}{E \vdash e \sqcup e' \hookrightarrow \top \text{ and } E \vdash e \sqcap e' \hookrightarrow \bot} \qquad \frac{E \vdash e \hookrightarrow 1[\bot] \text{ and } E \vdash e' \hookrightarrow \bot[1]}{E \vdash e \sqcup e' \hookrightarrow 1 \; and \; E \vdash e \sqcap e' \hookrightarrow \bot}$$

$$\frac{E \vdash e \hookrightarrow 0[\bot] \text{ and } E \vdash e' \hookrightarrow \bot[0]}{E \vdash e \sqcup e' \hookrightarrow 0} \qquad \frac{E \vdash e \hookrightarrow 0[\top] \text{ and } E \vdash e' \hookrightarrow \top[0]}{E \vdash e \sqcap e' \hookrightarrow 0}$$

$$\frac{E \vdash e \hookrightarrow x \text{ and } E \vdash e' \hookrightarrow x (x = \bot, 0, 1, \top) E \vdash e \hookrightarrow 1[T] \text{ and } E \vdash e' \hookrightarrow T[1]}{E \vdash e \sqcup e' \hookrightarrow x \text{ and } E \vdash e \sqcap e' \hookrightarrow x \qquad E \vdash e \sqcap e' \hookrightarrow 1}$$

of the respective output environments of P_1 and P_2. The parallel is ready to leave when both P_1 and P_2 are. The rule for $\|$ is:

$$\langle P_1 \rangle_E \sqcup \langle P_2 \rangle_E \vdash \mathcal{C}(P_1) \cup \mathcal{C}(P_2) \cup \mathcal{C}_{P_1 \| P_2} \hookrightarrow \langle P_1 \| P_2 \rangle_E$$

where

$$\mathcal{C}_{P_1 \| P_2} = \begin{array}{ll} \text{Set}_{P_1} = & \text{Set}_{P_1 \| P_2} \\ \text{Set}_{P_2} = & \text{Set}_{P_1 \| P_2} \\ \text{Reset}_{P_1} = & \text{Reset}_{P_1 \| P_2} \\ \text{Reset}_{P_2} = & \text{Reset}_{P_1 \| P_2} \\ \text{R}_1{}^+ = & (\text{RTL}_{P_1} \sqcup \text{R}_1) \sqcap \neg\text{Reset}_{P_1 \| P_2} \\ \text{R}_2{}^+ = & \text{RTL}_{P_2} \sqcup \text{R}_2) \sqcap \neg\text{Reset}_{P_1 \| P_2} \\ \text{RTL}_{P_1 \| P_2} = & (\text{RTL}_{P_1} \sqcup \text{R}_1) \sqcap (\text{RTL}_{P_2} \sqcup \text{R}_2) \end{array}$$

Notice that the circuit associated with the parallel operator requires two registers R_1 and R_2 to record the respective RTL values of both sides, because the parallel operator is ready to leave when its both arguments are.

The equational semantics provide us with an operational means to compile LE programs. It associates a ξ-equation system to each LE term and the evaluation of this last is done with respect to a constructive propagation law. A program is causal when we can sort its equation system. The equational semantics is responsive and modular. Then, this semantics is not causal. Although, we can rely on it to perform separated compilation of programs, We define a new sorting algorithm (based on the well-known PERT [18] technique) allowing to sort an equation system from already sorted sub parts. To complete our approach, we have shown that the two semantics presented in this paper coincide. For each reaction, given an input environment, if the program is causal, the equation system computes some output values and then the behavioral semantics computes identical outputs.

Theorem 1. *Let P be a* LE *statement, O its output signal set and E_C an input environment, the following property holds:*

$$\Gamma(P) \xrightarrow[E]{E',\mathrm{RTL}(P)} \Gamma(P') \text{ and } \langle P\rangle_{E_C}\lceil_O = E'\lceil_O$$

where $E\lceil_X$ means the restriction of environment to the the events associated with signals in X. $E = \{S^x | S^x \in E_C \text{ and } S \notin \mathcal{W} \cup \mathcal{R}\}$. RTL_P (RTL wire of $\mathcal{C}(P)$) can be considered as a boolean because it cannot be evaluated to $\bot$ or $\top$ and its value is the termination flag of behavioral semantics.

3 LE Modular Compilation

3.1 Introduction

In the previous section, we have shown that every construct of the language has a semantics expressed as a set of ξ equations. The first compilation step is the generation of a ξ equation system for each LE program, according to the semantics laws described in section 2.2.

But to rely on equation systems to generate code, simulate or link with external code requires to find an evaluation order, valid for all synchronous instants. Usually, in the most popular synchronous languages existing, this order is static. This static order forbids any separated compilation mechanism as shown in the example in figure 1.

To avoid such a drawback, independent signals must stay not related : we aim at building an incremental partial order. Hence, we keep enough information on signal causality to preserve the independence of signals. At this aim, we define two integer variables for each equation, namely (*CanDate*, *MustDate*) to record the date when the equation is evaluated. The CanDate characterizes the earliest date when the equation *can* be evaluated. The MustDate characterizes the latest date the equation *must* be evaluated to respect the critical time path of the equation system. Dates are ranges in a discrete time scale and we call them *levels*. Level 0 characterizes the equations evaluated first because they depend of free variables, while level n+1 characterizes the equations that require the evaluation of variables from lower levels (from n to 0) to be evaluated. Equations of same level are independent and so can be evaluated whatever the chosen order is. In fact, a couple of levels is associated with each equations. This methodology is inspired from the PERT method [18]. This latter is well known for decades in

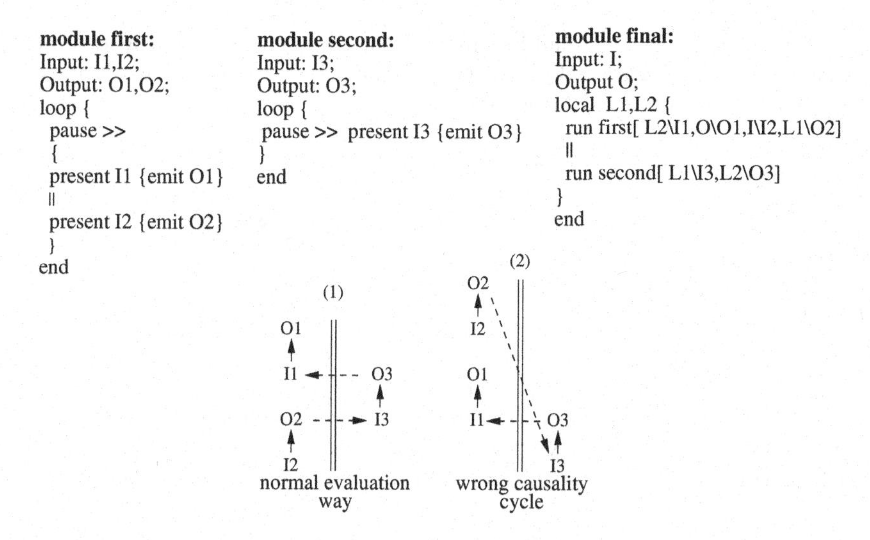

Fig. 1. Causality cycle generation. The **pause** instruction waits an instant to avoid instantaneous loop in modules **first** and **second**. In this example, O1, O2 and O3 signals are independent. Module first equation system is {O1 = I1, O2 = I2} and module second has { O3 = I3 } as equation system. But, when choosing a total order, we can introduce a causality cycle. If ordering (1) is chosen, in module final, taking into account the renaming, we obtain the system: { L1 = I, L2 = L1, O = L2 } which is well sorted. At the opposite, if we choose ordering (2), in module final we get: { L2 = L1, O = L2,L1 = I } which has a causality cycle.

the industrial production. This technique allows to deal with partial orders in equation systems and not immediately force the choice of a total arbitrary order.

3.2 Sort Algorithm: A PERT Family Method

Our sorting algorithm is divided into two phases. The first step constructs a forest where each tree represents variable dependencies. Thus initial partial orders sets are built. The second step is the recursive propagation of *CanDate* and *MustDate*. If during the propagation, a cycle is found there is a causality cycle in the program. Of course the propagation ends since the number of variables is finite. At worst, if the algorithm is successful (no causality cycle is found), we can find a total order with a single variable per level (n equations and n levels).

Sorting algorithm Description

More precisely, the first step builds two dependency sets (*upstream, downstream*) for each variable with respect to its associated equation. The *upstream* set of a variable x is composed of the variables needed by x to be computed while the *downstream* set is the variables that need the value of x to be evaluated. The toy example described in table 3 illustrates how we build upstream and downstream

Table 3. An equation system and its upstream and downstream dependency graphs. the CanDate and MustDate table denotes the computed dates of each variable, for instance a has CanDate 1 and MustDate 1).

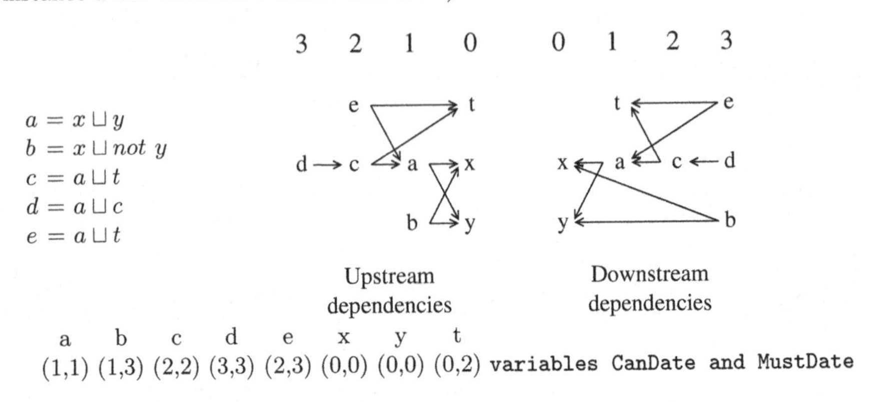

partial orders sets. First, from equation system, we build two dependency graphs. Each graph leads to construct two partial orders sets. For instance, in table 3, the upstream partial order set contains: $\{d \rightarrow c \rightarrow a \rightarrow x\}$, $\{d \rightarrow c \rightarrow a \rightarrow y\}$, $\{b \rightarrow x\}$, $\{b \rightarrow y\}$, etc....

After the construction of variable dependencies, we perform *CanDate* and *MustDate* propagation. Initially, all variables are considered independent and their dates (*CanDate* , *MustDate*) are set to (0,0). The second step recursively propagates the *CanDate* according to the upstream partial orders set while the *MustDate* is propagated according to the downstream partial orders set. In table 3, the CanDate and MustDate of each variable is shown. As a result, we have several orders acceptable. Looking at the example (see table 3), there are four levels (from 0 to 3). At level 0, we have x, y which are free variables. Equations for c is at level 2 and equation for d is at level 3 while equation for b can be at each level between 1 and 3. Equation for e can be evaluated either at level 2 or 3 and its order against c or d equations is free. In practice, the propagation algorithm we implement has a $n\ log\ n$ complexity.

Link of two Partial Orders

The approach allows an efficient link of two already sorted equation systems, to perform separated compilation, for instance. We don't need to launch the sorting algorithm from its initial step.

To link two equation systems, we only consider their common variables. In fact, as a consequence of equational semantics we rely on, we need to link an output variable of a system with an input one of the other and conversely. Assume that an output variable x of a sorted equation system A is an input variable of an another sorted equation system B. Let us denote $C^A(x)$ (resp $C^B(x)$) the CanDate of x in A (resp B). Similarly, we will denote $M^A(x)$ (resp $M^B(x)$) the MustDate of x in A (resp B). Then, we compute $\Delta_c(x) = |C^A(x) - C^B(x)|$, thus we shift by $\Delta_c(x)$ the CanDate of variables that depend of x in both equation systems (we look at the respective upstream partial orders computed for A and B equation systems). Similarly, we compute $\Delta_m(x) = |M^A(x) - M^B(x)|$, to shift the MustDate of variables that need x looking at the respective downstream partial orders of both equation systems.

3.3 Practical Issues

We have mainly detailed the theoretical aspect of our approach, and in this section we will discuss the practical issues we have implemented.

Effective compilation

We rely on the equational semantics to compile LE programs. To this aim we first compute the equation system associated with the body of a program (by computing the circuit associated with each node of the syntactic tree of the program). Doing that, according to the constructive approach we trust in, we also refine the value of signal and register status in the environment. Then, we translate each ξ circuit into a boolean circuit. To achieve this translation we define a mapping for encoding elements of ξ-algebra with a pair of boolean values. This encoding allows us to translate ξ equation system into a boolean equation system (each equation being encoded by two boolean equations). Finally, the effective compilation turns out to be the implementation of $\hookrightarrow$ propagation law to compute output and register values.

We call the compilation tool that achieves such a task CLEM (Compilation of LE Module). In order to perform separate compilation of LE programs, we define an internal compilation format called LEC (LE Compiled code). This format is highly inspired from the Berkeley Logic Interchange Format (BLIF[3]). This latter is a very compact format to represent netlists and we just add to it syntactic means to record the *CanDate* and *MustDate* of each variable defined by an equation. Practically, CLEM compiler, among other output codes, generates LEC format in order to reuse already compiled code in an efficient way (see section 3.2), thanks to the PERT method we implement.

Finalization Process

This approach to compile LE programs into a sorted ξ equation system in an efficient way requires to be completed by what we call a *finalization* phase. This latter is specific to our approach because the separated compilation implies to keep the $\perp$ status of signals for which no information has been propagated during the compilation phase, unlike the others synchronous approaches. A classical difficult problem in the compilation of synchronous languages is the "reaction to absence": signals are never set to absent and their final absent status is detected

[3] `http://embedded.eecs.berkeley.edu/Research/vis`

by a global propagation of signal status on the overall program. But these compilation algorithms prevent any separated compilation because they need a global vision of all status signals. In our approach, we keep the $\perp$ status of signal until the end of compilation and we drive back the absence resolution at the end of the process and we call it finalization. Hence, we replace all $\perp$ events by absent events. Notice that the finalization operation is harmless. The sorting algorithm relies on propagation of signal values, and the substitution of $\perp$ by 0 cannot change the resulting sorted environment.

Compilation scheme

Now, we detail the workflow we have implemented to specify, compile , simulate and execute LE programs. LE language help us to design programs. In the case of automaton, it can be generated by automaton editor like `galaxy` too. Each LE module is compiled in a LEC file and includes instances of RUN module references. These references can have been already compiled in the past by a previous call to the `clem` compiler. When the compilation phase is achieved, the finalization will simplify the final equation system and generates a file for a targeted usage: simulation, hardware description or software code. As mentioned in the introduction, an attractive consequence of our approach is the ability to perform validation of program behaviors. As a consequence, we can rely on model checking based tools to verify property of LE language. Thus, from lec files, we also generate `smv` files to feed the NuSMV model checker [4]. The `clem` workflow is

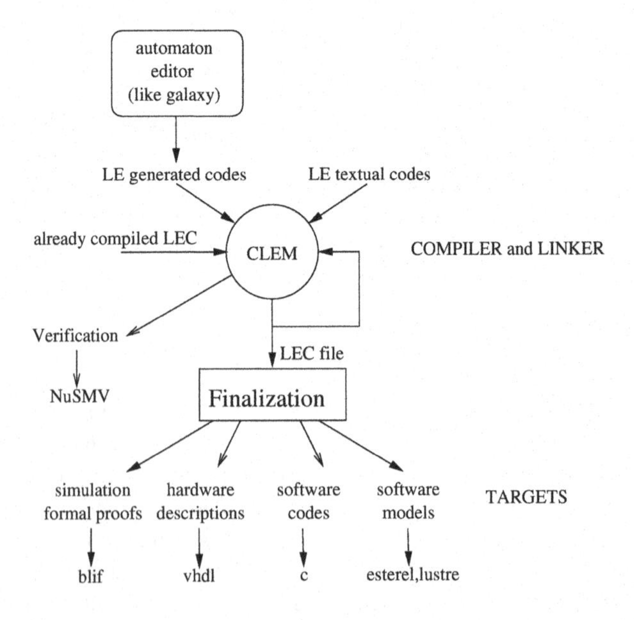

Fig. 2. Compilation Scheme

summed up in the figure 2. All softwares belonging to the toolkit are available at: "http://www.unice.fr/L-EEA/gaffe/LE.html".

4 Related Works

The motivation for this work is the need for research on efficient and modular compilation of synchronous programs. Model-driven languages have been advocated to allow reliable and reusable system design. But rely on a responsive and modular semantics to compile a synchronous language is not at all straightforward since the causality problem must be faced. In [7], S Edwards and E Lee introduce a block diagram language to assemble different kinds of synchronous software. They introduce a fixpoint semantics inspired by the Esterel constructive semantics and remain deterministic even in the presence of instantaneous feedback. They propose exact and heuristic algorithms for finding schedules that minimize system execution time. But their block diagram language is an assembly language without hierarchy, while [13], consider hierarchical block diagrams. The purpose of this work is to generate code for "macros" (i.e a block diagram made of sub blocks) together with a set of interface functions in order to link with an external context. This idea is close to ours, but block diagrams are made of a hierarchy of sub diagrams interconnected and have not the expressiveness of LE language. Moreover, we can reload already compiled file dynamically in our compilation process. In [20], the authors consider a partial evaluation of Esterel programs: they generate distributed code that tries to compute as many outputs as possible while some inputs are unknown. In [16], a synchronous language Quartz is introduced and its compilation into a target job language. The separated compilation is done by splitting a program into sequential jobs corresponding to control flow locations of the program. This approach is well suited to generate distributed software. Similarly, some attempts allow a distributed compilation of programs [19, 6] but they don't address the problem of dynamic link of already compiled sub programs.

5 Conclusion and Future Work

In this work, we introduced a synchronous language LE that supports a separated compilation. We defined its behavioral semantics giving a meaning to each program and allowing us to rely on formal methods to achieve verification. Then, we also defined an equational semantics to get a means to really compile programs in a separated way. Actually, we have implemented the `clem` compiler. This compiler is a link in the design chain we have to specify control-dominated process from different front-ends: a graphical editor devoted to automata drawing, or direct LE language specification to several families of back-ends.

In the future, we plan to improve our approach. The first improvement we aim at, is the extension of the language. To be able to deal with control-dominated systems with data (like sensor handling facilities), we will extend the syntax of the language. Then, we plan to integrate abstract interpretation techniques (like polyhedra intersection, among others) [5] to take into account data constraints in control. Moreover, we also need to discuss with signal processing or automation world through their specific tool Matlab/Simulink (http://www.mathworks.com). Another major improvement we are interesting in, concerns the development of verification means. Synchronous approach provides us with well-suited models to apply model checking techniques to LE programs. We already connect to the NuSMV model checker and provide with symbolic and bounded model-checking techniques application. A verification means successfully used for synchronous formalisms is that of observer monitoring [9]. We plan to introduce the ability to define safety properties as observers in LE and internally call NuSMV to prove them. Moreover, our modular approach opens new ways to perform modular model-checking. We need to prove that "assume-guarantee" technique [12] applies in our formalism.

References

1. André, C., Boufaïed, H., Dissoubray, S.: Syncharts: un modéle graphique synchrone pour systéme réactifs complexes. In: Real-Time Systems (RTS 1998), Paris, France, January 1998, pp. 175–196. Teknea (1998)
2. Berry, G.: The Constructive Semantics of Pure Esterel. Draft Book (1996), `http://www.esterel-technologies.com`
3. Berry, G.: The Foundations of Esterel. In: Plotkin, G., Stearling, C., Tofte, M. (eds.) Proof, Language, and Interaction, Essays in Honor of Robin Milner. MIT Press, Cambridge (2000)
4. Cimatti, A., Clarke, E., Giunchiglia, E., Giunchiglia, F., Pistore, M., Roveri, M., Sebastiani, R., Tacchella, A.: NuSMV 2: an OpenSource Tool for Symbolic Model Checking. In: Brinksma, E., Larsen, K.G. (eds.) CAV 2002. LNCS, vol. 2404, pp. 359–364. Springer, Heidelberg (2002)
5. Cousot, P., Cousot, R.: On Abstraction in Software Verification. In: Brinksma, E., Larsen, K.G. (eds.) CAV 2002. LNCS, vol. 2404, p. 37, 56. Springer, Heidelberg (2002)
6. Edwards, S.A.: Compiling esterel into sequential code. In: Proceedings of the 7th International Workshop on Hardware/Software Codesign (CODES 1999), Rome, Italy, May 1999, pp. 147–151 (1999)
7. Edwards, S.A., Lee, E.A.: The semantics and execution of a synchronous block-diagram language. Science of Computer Programming 48(1), 21–42 (2003)
8. Halbwachs, N.: Synchronous Programming of Reactive Systems. Kluwer Academic, Dordrecht (1993)
9. Halbwachs, N., Lagnier, F., Raymond, P.: Synchronous observers and the verification of reactive systems. In: Nivat, M., Rattray, C., Rus, T., Scollo, G. (eds.) Third Int. Conf. on Algebraic Methodology and Software Technology, AMAST 1993, Workshops in Computing, Twente, June 1993. Springer, Heidelberg (1993)
10. Harel, D., Pnueli, A.: On the development of reactive systems. In: NATO, Advanced Study institute on Logics and Models for Verification and Specification of Concurrent Systems. Springer, Heidelberg (1985)
11. Huizing, C., Gerth, R.: Semantics of reactive systems in abstract time. In: de Roever, W.P., Rozenberg, G. (eds.) Real Time: Theory in Practice, Proc. of REX workshop, June 1991. LNCS, pp. 291–314. Springer, Heidelberg (1991)

12. Clarke Jr., E.M., Grumberg, O., Peled, D.: Model Checking. MIT Press, Cambridge (2000)
13. Lublinerman, R., Tripakis, S.: Modularity vs. reusability: Code generation from synchronous block diagrams. In: Design, Automation and Test in Europe (DATE 2008) (2008)
14. Potop-Butucaru, D., De Simone, R.: Optimizations for Faster Execution of Esterel Programs. In: Gupta, R., LeGuernic, P., Shukla, S., Talpin, J.-P. (eds.) Formal Methods and Models for System Design, Kluwer, Dordrecht (2004)
15. Ressouche, A., Gaffé, D., Roy, V.: Modular compilation of a synchronous language. Research Report 6424, INRIA (January 2008), `https://hal.inria.fr/inria-00213472`
16. K. Schneider, J. Brand, E. Vecchié. Modular compilation of synchronous programs. In *From Model-Driven Design to Resource Management for Distributed Embedded Systems*, volume 225 of *IFIP International Federation for Information Processing*, pages 75–84. Springer Boston, January 2006.
17. Esterel Technologies. Esterel studio suite, `http://www.esterel-technologies.com`
18. Kirkpatrick, T.I., Clark, N.R.: Pert and an aid to logic design. IBM Journal of Research and Develepment, 135–141 (March 1966)
19. Weil, D., Bertin, V., Closse, E., Poize, M., Venier, P., Pulou, J.: Efficient compilation of esterel for real-time embedded systems. In: Proceedings of the 2000 International Conference on Compilers, Architecture, and Synthesis for Embedded Systems, San Jose, California, United States, November 2000, pp. 2–8 (2000)
20. Zeng, J., Edwards, S.A.: Separate compilation for synchronous modules. In: Yang, L.T., Zhou, X.-s., Zhao, W., Wu, Z., Zhu, Y., Lin, M. (eds.) ICESS 2005. LNCS, vol. 3820, pp. 129–140. Springer, Heidelberg (2005)

Describing Active Services for Publication and Discovery

Haldor Samset and Rolv Bræk

Norwegian University of Science and Technology (NTNU), Dept. of Telematics,
N-7491 Trondheim, Norway
haldors@item.ntnu.no, rolv@item.ntnu.no

Summary. Services in contemporary SOA are typically considered to be of passive nature, providing functionality that solely executes upon invocation, while active services contain functionality that cannot be limited to initiative in a one direction only. We have earlier introduced the notion of active services in the context of SOA, concentrating on modeling these using UML 2 Collaborations. Here we suggest how active services and their behavior can be described for publication and discovery in the context of SOA, focusing on the service description as a behavioral contract.

1 Introduction

The service-oriented architecture (SOA) is essentially an architectural style where program functionality is logically organized into services that are universally accessible through well-defined interfaces, supported by mechanisms to publish and discover the available functionality and the means of communicating with the service. Ideally, a service should provide a rather self-sufficient piece of functionality applicable within a particular context. The organization of functionality into services is meant to give rise to a larger degree of functionality reuse. This is achieved by keeping service interfaces independent from functionality implementations, keeping them independent from other service interfaces and by enabling universal access.

Services in contemporary[1] SOAs are inherently passive due to the pervading client-server paradigm. Communication initiative is always taken by the client side, and the server side is reduced to answer these requests. We have in an earlier paper [21] investigated the notion of active services, and argued that services in a SOA should allow for functionality that cannot be limited to initiative in one direction only.

In this paper we continue to investigate the application of active services by dealing with publication and discovery of such services in SOA. We start by quickly recapping the notion of active services and how they are modeled in section 2. Then we continue with examining how active services can be described

[1] As Erl[5], we use *contemporary SOA* to refer to the typical WSDL- and Web Services-based service-oriented architectures usual in the industry today.

R. Lee (Ed.): Soft. Eng. Research, Management & Applications, SCI 150, pp. 173–187, 2008.
springerlink.com

for publication in section 3, and section 4 illustrates how the published service descriptions can be applied in a SOA. Finally we take a look at some related work in section 5 before we sum up and present an outlook on further work in section 6 .

2 Active Services

We consider the typical services in contemporary SOAs to be passive, in the sense that the functionality they provide is considered to be executed solely upon invocation. In this context, the term *service request* is typically applied to describe the invocation of an operation available in a declared interface of a service. The service concept in SOA has grown out of the Web Service technology, and is tightly tied to the concept of interface as declared by WSDL, which is restricted to describing the static invocation details of interfaces. In this sense, the client-server paradigm of HTTP pervades the service model of SOA.

However, a rich class of services are of an active nature, in the sense that their functionality is not only invoked by an client, but is active behavior that may take initiative towards the client, seemingly on their own. The classic telephone service is a good case in point as a user may both initiate and receive calls. Likewise, the numerous scenarios describing advanced context-aware services performing proactive behavior exemplify active service functionality that is not inherently passive. In general, services involving several active participants (humans or devices) that may take independent initiatives to interact are of an active nature.

The participants of active services cannot well be classified as "clients" or "servers", and thus they do not fit naturally into the client-server paradigm of contemporary SOA. It is possible to design workarounds for this mismatch, but they invariably add complexity for the service design. Active objects with concurrent behavior do not communicate well by invocation, since this involves transfer of control, and blocking behavior[10]. Some sort of signal sending or messaging using buffered communication is better suited for active service interactions.

2.1 Services and State

Frequently encountered in the SOA literature is the statement that *services should be stateless* (Erl, for instance, lists this among his service-orientation principles [5]). We argued in [21] that many entities entail a stateful session behavior that really matters for clients, and that it should be represented clearly to handle the complexities in a controlled way. Our service engineering approach promotes explicit modeling of individual service sessions and their behavior. If the behavior happens to be stateless, so much the better. If it is stateful, it should be made explicit as this is essential to validate interfaces and manage service composition. Consequently, it is necessary to factor out the individual service sessions and model these explicitly, considering how the sessions are mapped on the physical entities and interfaces as a separate issue.

2.2 Modeling Active Services

Let us explain how we model active services by looking at an example. Notification of incoming SMS messages is a common type of functionality that is offered by telecom providers to third parties. Mobile phone users can send messages (often including special keywords) to a special short number, and the telecom provider then informs the corresponding third party when messages arrive. The Parlay X specification [11] specifies a collection of Web Services intended to make telecom functionality like call control and SMS available for third parties. In an earlier paper [21] we combined the services related to SMS notification into an example active service named SmsNotification. Rather than two separated services that were inevitably dependent on each other, the resulting active SmsNotification service nicely contains the functionality to start receiving notifications of incoming SMS messages and keep receiving them until ending the notification.

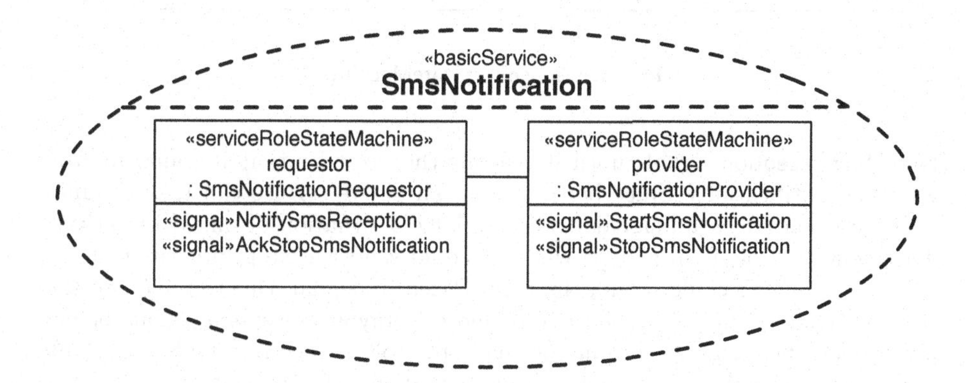

Fig. 1. The active SmsNotification service

Figure 1 shows the SmsNotification service. We use stereotyped UML 2 Collaborations (introduced in UML 2.0 [18]) when modeling active services. The BasicService stereotype constrains the collaboration to contain only two collaboration roles (requestor and provider), and each role is typed with an interface that states which signals that the role can receive. The visible behavior of the roles in a BasicService are described using state machines that conform to the ServiceRoleStateMachine stereotype, stating the order of sending and receiving signals. The ServiceRoleStateMachine is a very restricted behavioral state machine, containing a single region with exactly one initial state. Transitions are limited to have triggers of type SendSignalEvent or ReceiveSignalEvent, and transitions cannot specify any constraints (guards, preconditions, postconditions) [21].

Figure 2 shows the state machines for the requestor and provider roles of the SmsNotification service, where "!" denotes a SendSignalEvent and "?" a ReceiveSignalEvent. The requestor role requests to be notified of incoming SMS messages by sending a StartNotification signal to the provider, and it receives

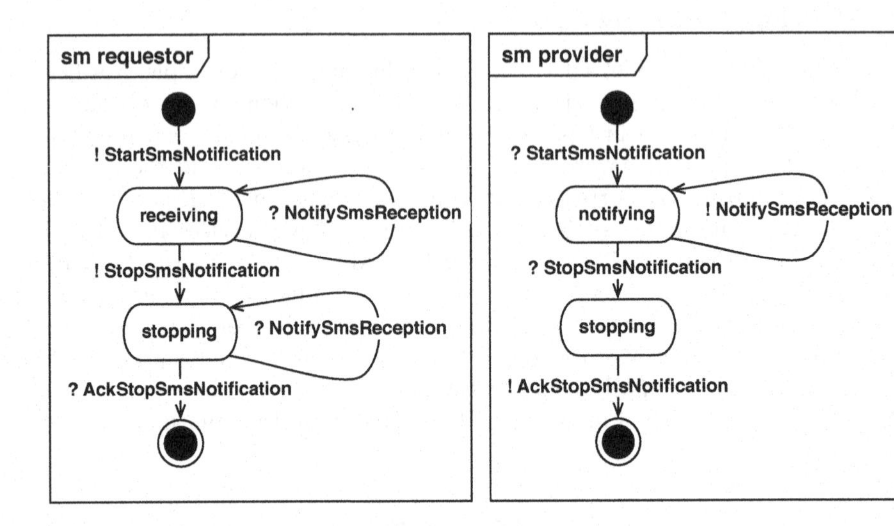

Fig. 2. ServiceRoleStateMachines

NotifySmsReception signals until it informs the provider that it wants to stop receiving notifications. Since both roles in an active service can take communicative initiative (send signals), we here have a situation of mixed initiatives. The requestor role is in a state where it could send the StopNotification signal while the provider could send a NotifySmsReception signal. Due to the delay, the provider could manage to send one or more NotifySmsReception signals before the StopNotification signal from the requestor role is received. One may think that losing a signal or two is not that important since the requestor is ending its behavior anyway. But this behavior is non-intended, and non-intended behavior may lead to logical errors in a system. In this case we also deal with a network resource (SMS messages) that in the end somebody has to pay for. Incoming SMS messages are either paid by the user that sent it, or the third party that uses the SmsNotification service (according to the usage terms with the SMS provider). Simply dropping messages paid for by the user is not good business practice.

The problem with the mixed initiatives is solved rather easily by making the requestor role wait for a AckStopNotification signal from the provider before ending the service interaction. Any NotifySmsReception signals already on their way are then received while waiting.

We have developed an Eclipse plugin for modeling active services of as UML 2 Collaborations in our integrated service engineering tool suite Ramses [13]. The editor is for now restricted to active services of type BasicServices, and includes an editor for modeling the ServiceRoleStateMachines. Another plugin generates a simple Promela model of the service interaction, thus enabling us to validate the behavior using the Spin model checker [12].

2.3 Collaborations as Service Contracts

One of the well-known "four tenets of service-orientation" by Don Box of Microsoft [3] is that "*services share schema and contract*". The idea is that a contract describes the structure and ordering constraints of the messages belonging to the service, and a schema describes the conceptual data manipulated. Box's tenets are widely cited in the SOA community, but the notion of contract has got little attention as most seem content with organizing operations into interfaces.

The two ServiceRoleStateMachines of a BasicService state exactly how the entities participating in the service are allowed and expected to behave. In our view, they form a precise contract for the service that is in terms with Box's tenets.

Note that the collaboration in Figure 1 does not detail the behavior of entities implementing the service, only the visible interface behaviors required by any pair of entities participating in the service. In this way it defines a *behavioral contract* that participants must obey. It can be bound to all entities satisfying its role behaviors and thus it is reusable as a contract and serves to facilitate reuse of the entities it is bound to.

Note that this kind of contract includes properties of the connection as well as the behavior required at both ends. This is important because the compatibility of behaviors to some extent depends on properties of the connection linking them; e.g. whether signals are ensured to be received in the same order as they are sent. Based on the specified role behaviors and the connecting link, one may validate the correctness of the contract separately from any entity is is bound to.

It is important to understand at this point that there are well defined mappings from entity behaviors, defined as state machines, to interface role behaviors [9].

Interface role behaviors can be automatically derived from the more complete entity behaviors. They can be compared with role behaviors specified in service contracts, and also be used to define new contracts.

3 Describing Active Services for Publication

What exactly do we need to describe in order to publish an active service using a contract? Central in the active service concept is the notion of roles and the signals communicated between these. A role in our model consists of signal receptions and a behavioral state machine of the ServiceRoleStateMachine stereotype. At a minimum, an active service description should describe which signals the roles can receive, as well as the syntactical details of these signals. This corresponds to the typical interface descriptions of WSDL, which is limited to describe static syntactical and invocation details.

Platform-independent interface descriptions using well-established standards is the key factor for a service-oriented architecture. XML Schema Definition (XSD) [7] is the de facto standard for describing data structures in contemporary SOAs. Hence, XSD a natural choice for describing the signals and their parameters. Figure 3 demonstrates how the signals used in the SmsNotification service are expressed using XSD. (In the figure only the NotifySmsReception

```
<xsd:schema targetNamespace="http://no.ntnu.item/smsnotification.xsd">
  <xsd:element name="StartSmsNotification"/>
  <xsd:element name="NotifySmsReception">
   <xsd:complexType>
    <xsd:sequence>
      <xsd:element name="message" type="xsd:string"/>
      <xsd:element name="senderAddress" type="xsd:anyURI"/>
      <xsd:element name="smsServiceActivationNumber" type="xsd:anyURI"/>
      <xsd:element name="dateTime" type="xsd:dateTime"/>
    </xsd:sequence>
   </xsd:complexType>
  </xsd:element>
  <xsd:element name="StopSmsNotifcation"/>
  <xsd:element name="AckStopSmsNotification"/>
 </xsd:schema>
```

Fig. 3. Signal types expressed using XSD

signal is described in full with detailed parameters, the remaining signals are presented with the top element only for brevity).

The BasicService contains precise behavioral descriptions of the service roles, stating the order of which signals could be sent and received. We find no reason not to include these descriptions when describing a service, to the contrary we believe that they are vital in understanding the service interaction. The ServiceRoleStateMachines are already represented using XML, due to the fact that the UML models are made using our plugins based on the UML 2 Eclipse plugins. The UML models in Eclipse are persisted using XMI, a XML format for exchanging UML models. However, XMI is mainly intended for storage and design-time use, with a focus on expressiveness. The ServiceRoleStateMachines in XMI format are quite verbose, containing for instance extensive details about trigger and event types. Consequently, a ServiceRoleStateMachine expressed in XMI is not especially suited for publication. We suggest a more condensed and compact XML format for the role behavior, expressing only the list of states, and the possible transitions between them resulting from the role either sending or receiving a signal.

Figure 4 shows the XML representation of the requestor role of the SmsNotification service. Studying at the description, we see that it starts with declaring the initial state (which is required for all ServiceRoleStateMachines). In this initial state, the only behavior allowed by the requestor role is to send the StartSmsNotification signal. When this signal is sent, it enters the state named "receiving". In this state, the requestor it can either receive a NotifySmsReception signal, or it can send the StopSmsNotification signal to the provider. In the first case, the requestor reenters the "receiving" state and is once more able to perform one of the actions. In the latter case, it enters the state "stopping" where it waits for the AckStopSmsNotification signal that confirms that the provider has stopped

```xml
<rolesm type="SmsNotificationRequestor">
 <state id="initial">
  <send signal="StartSmsNotification" nextState="receiving"/>
 </state>
 <state id="receiving">
  <receive signal="NotifySmsReception" nextState="receiving"/>
  <send signal="StopSmsNotification" nextState="stopping"/>
 </state>
 <state id="stopping">
  <receive signal="NotifySmsReception" nextState="stopping"/>
  <receive signal="AckStopSmsNotification" nextState="final"/>
 </state>
</rolesm>
```

Fig. 4. ServiceRoleStateMachine as XML

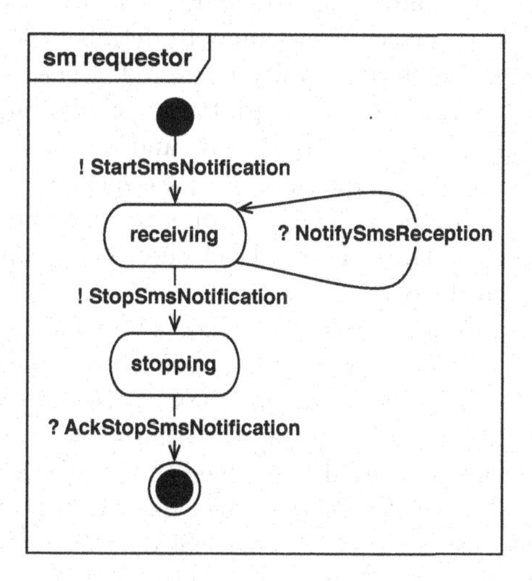

Fig. 5. Faulty mirrored requestor role

sending signals. The requestor can also receive any NotifySmsReception signals that may be sent from the provider before it received the request to stop the notification, thereby ensuring that no NotifySmsReception signals are lost. Upon receiving the AckStopSmsNotification signal, the requestor ends its behavior by entering the final state.

Contrary to the client-server type services common in contemporary SOAs, our active service concept consists of two roles; the requestor and the provider.

We have stated that an entity in need of a certain functionality typically would fill the requestor role, while the provider role represents the entity that offers this functionality. It may therefore seem enough to publish only the description of the provider role. After all, the required requestor interface is given indirectly through which signals the provider role can send. Likewise, the ServiceRoleStateMachine of a requestor can be constructed based on the provider role. Note that mixed initiatives that is not resolved by the provider must be acknowledged and resolved during this construction. For instance, directly mirroring the provider in the SmsNotification example would produce a ServiceRoleStateMachine as shown in 5. This state machine does not anticipate any NotifySmsReception signals that could be on their way in the "stopping" state, despite waiting for the AckStopNotification signal. As we already have a correct "mirror" of the provider role available through the requestor role, we see no reason not to include the description of it as well.

3.1 Describing Active Services with WSDL

The Web Services Description Language (WSDL) is, as the name indicates, a XML language for describing Web Services. According to the specification of WSDL version 2.0 [16], it "enables one to separate the description of the abstract functionality offered by a service from concrete details of a service description such as how and where that functionality is offered". The actual description is divided into an abstract and a concrete part. At the abstract level, the service is described "in terms of the messages it sends and receives". The messages are described as types using XSD. One or more messages are used in conjunction with a message exchange pattern to form an operation, and operations are again grouped into an interface. The concrete level describes bindings, which specify transport details for interfaces.

Apart from the focus on operations, WSDL 2.0 sounds somewhat adequate for our service descriptions. We have already decided to describe the signals using XSD, and WSDL 2.0 allow us to import these directly into the types declaration. Due to the extensibility of WSDL, the XML descriptions of the ServiceRoleStateMachines could easily be added to the WSDL document. WSDL 2.0 introduces eight Message Exchange Patterns (MEPs) [19, 15] that can be used for defining the directions of messages in operations. With the out-only and in-only MEPs we can describe which signals a role sends and receives when we declare the abstract interface of a role. Figure 6 show the abstract interface declaration of the SmsNotificationProvider interface using WSDL 2.0. Notice the "pattern" attribute on the operations, which states the applied MEP. All operations in the declared interface apply either *in-only* (for signal receptions) or *out-only* (for signal sending). The use of MEPs may seem somewhat superfluous when we include the XML ServiceRoleStateMachine descriptions, but they are in terms with the WSDL 2.0 specification. This could be useful for requesting entities that cannot make use of our ServiceRoleStateMachine descriptions.

We have now covered the abstract part of the WSDL 2.0 description, but what about the concrete part? What is missing is the declaration of a binding

```
<types>
  <xsd:import namespace="http://no.ntnu.item/smsnotification"
      schemaLocation="http://no.ntnu.item/smsnotification.xsd"/>
</types>
<interface name="SmsNotificationProvider">
  <operation name="receiveStartSmsNotification"
      pattern="http://www.w3.org/ns/wsdl/in-only">
    <input element="StartSmsNotification" messageLabel="In"/>
  </operation>
  <operation name="sendNotifySmsReception"
      pattern="http://www.w3.org/ns/wsdl/out-only">
    <output element="NotifySmsReception" messageLabel="Out"/>
  </operation>
  <operation name="receiveStopSmsNotification"
      pattern="http://www.w3.org/ns/wsdl/in-only">
    <input element="StopSmsNotification" messageLabel="In"/>
  </operation>
  <operation name="sendAckStopSmsNotification"
      pattern="http://www.w3.org/ns/wsdl/out-only">
    <output element="AckStopSmsNotification" messageLabel="Out"/>
  </operation>
</interface>
```

Fig. 6. WSDL abstract interface

of the abstract interface, which explains how to access a service in terms of how the messages are exchanged. A binding should according to the standard specify "concrete message format and transmission protocol details for an interface" [16]. The WSDL 2.0 specification declares bindings for SOAP and HTTP that opens for extending with bindings for other technologies. A binding should include binding details for each operation of an interface, but the specification also define some defaulting rules [19]. Finally, we need to define a service (in WSDL 2.0 terms), which declares where an implementation of the declared binding can be accessed . This is done with a list of endpoint locations where that interface can be accessed, and the endpoints must reference one of the defined bindings in order to indicated the protocols and transmission formats that are to be used. Figure 7 shows the binding and service declaration in WSDL 2.0 for a SOAP implementation of the SmsNotificationProvider interface (note that the applied SOAP MEP *request-response* is not an error, but the correct MEP when binding a in-only operation using SOAP).

Our system engineering approach is based on a meta-model for design and execution of services. Central in this model is precise behavioral descriptions of components as communicating extended finite state machines in the form

```
<binding interface="SmsNotificationProvider"
         name="SmsNotificationProviderSoapBinding"
         type="http://www.w3.org/ns/wsdl/soap"
         mepDefault="http://www.w3.org/2003/05/soap/mep/request-response"
         protocol="http://www.w3.org/2003/05/soap/bindings/HTTP/">
  <operation ref="StartSmsNotification"/>
  <operation ref="StopSmsNotification"/>
</binding>
<service interface="SmsNotificationProvider" name="SmsNotificationService">
   <endpoint address="http://www.item.ntnu.no/examples/smsnotification/"
            binding="SmsNotificationProviderSoapBinding"
            name="SmsNotificationServiceSoapEndpoint"/>
</service>
```

Fig. 7. SOAP binding

of UML 2.0 state machines. The execution model relies on asynchronous communication between components using signal buffers. This ensures a decoupled component model well suited for modeling and realization of distributed systems (see [14] for a detailed look on our execution model). An execution environment could apply SOAP for the signal transmission and still fulfill the properties of our execution model, but this requires that the buffered communication is ensured. Hence, all SOAP calls must be in-only, and the invocation should return immediately (typically after triggering an internal event at the invoked SOAP operation provider).

4 Application Example

Figure 8 describes an example SOA system where an application makes use of several active services that are published and discovered. The Treasure Hunt Application is an application that lets mobile phone users carry out a sort of treasure hunt, involving simple text puzzles. The user starts the treasure hunt by sending a SMS message to a specified short number. As a result of this, the Treasure Hunt Application receives this incoming SMS message through its participation in the SmsNotification service. The TrackLocation service provides tracking of the location of GSM terminals, hence requestors can subscribe to location updates of mobile phone users (presuming that they have allowed location tracking of their phone). The Treasure Hunt Application starts tracking the location of the user through the TrackLocation service. Based on the location of the user, the Treasure Hunt Application issues a puzzle where the solution is the name of a location within reach of the user. The user has to solve the puzzle in order to unveil the location, and has to get to this location within a specified period of time. If the user is unable to solve the puzzle, or does not get to the location in time, the treasure hunt is over and a SMS message is sent to the user by using

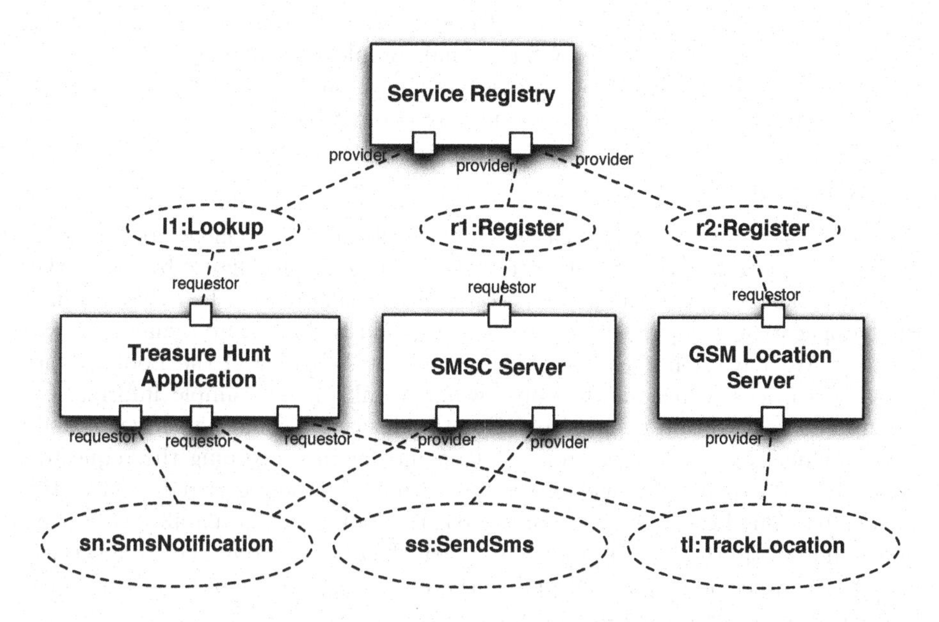

Fig. 8. The Treasure Hunt System

the SendSms service. If the user arrives at the location in time, the Treasure Hunt Application will detect this by location updates received through the TrackLocation service. A new puzzle is then sent out to the user, and the treasure hunt continues until a specific number of puzzles are solved satisfactory, or the user has failed to solve a puzzle.

Figure 8 serves to illustrate essential aspects of our approach. Treasure Hunt Application, ServiceRegistry, SMSC Server and GSM Location Server are distributed components, each assumed to have a complete behavior defined in terms of state machines, called the component behavior in the following. The system is service-oriented in the sense that these components interact with each other through participating in several active services, either by providing or requesting certain functionalities. The services are represented in the figure by UML CollaborationUses, each collaboration use representing a contract with two collaboration roles that define the visible interface behavior required by any component playing the roles. The compatibility among the collaboration roles are checked separately for each type of service contract using model checking and other techniques. The computational complexity of doing this is limited and need only be performed once for each contract type.

The collaboration roles of the service contracts are bound to the components of the treasure hunt system, as shown in Figure 8. This means that each component behavior must be compatible with all the role behaviors bound to the component. This is checked for each component separately by projecting its component behavior to the ports and comparing the projected port behaviors with the collaboration role behavior bound to the port. Since this can be done locally for each component type, the computational complexity is limited.

The two steps outlined above provide highly scalable mechanisms to ensure that all the links among the components are compatible.

4.1 Discovery

The services required by the Treasure Hunt Application component are obtained by looking them up in the ServiceRegistry, illustrated in the figure by use of the offered Lookup service. This lookup could either take place when the Treasure Hunt Application is deployed, or the services could be looked up dynamically at runtime. We have deliberately not described the behavior of the Lookup and Register services, which most people would think of as a simple information lookup and registration respectively. But these services could be active as well: The provider role of the Register service could for instance ping the requestor regularly and expect acknowledgments in return that the registered service still is available. The Lookup service could treat the lookups as availability subscriptions, and inform the Lookup requestor when the obtained reference to a service provider instance no longer is available, thereby triggering a new lookup in order to discover another suiting instance.

As mentioned earlier, we assert that the components have been validated against the visible interface behavior required by them in order to fulfill the roles of the services they are to participate in. In other words, that they will fulfill the published behavioral contract.

In the case where one is to design and implement a component from scratch, the published ServiceRoleStateMachine of the services that the component is going to request would be useful. The requestor role description could function as a blueprint for design and implementation, even though the developer does not have access to the UML model of the published services. Output and input signals of the component could be validated against the requestor role state machine of the services it will request. This validation could also be performed on existing components, and would be especially valuable in the cases where a published service has been extended, but still is supposed to work with a existing component originally designed to use an older version of that service.

5 Related Work

The notion that services can be of active nature is not new. It is more or less taken for granted in Vissers and Logrippo's classic paper about the service concept [23], where they define a service as "the behavior of the provider as observed by its users". The provider behavior is seen as an abstract machine which service primitives can be invoked upon, but it is also capable of spontaneous actions resulting in the execution of service primitives upon the user. Our use of state machines to describe service role behavior somewhat resembles this, but we also describe the interaction protocol of the service, which corresponds to the "internals of the service provider" in [23].

COSMO[20] is a proposed conceptual framework for service modeling and refinement, aimed at providing a common semantic meta-model for service languages. COSMO identifies two roles involved in a service; the user role and provider role, and our requestor and provider roles correspond well to these concepts. It seems that COSMO does not treat initiative, so whether initiative is restricted to the user role alone is not clear. The examples in [20] are all with client-server type initiatives, while our service concept are deliberately devised to allow peer-to-peer style initiatives. Causality relations between activities is the notion of state in COSMO, while we have chosen to deal with state explicitly with state machines when modeling the roles.

Mencl presented the *Port State Machine* (PSM) in [17] for modeling the communication on a port, which bears similarities with our ServiceRoleStateMachine, but they do not describe bidirectional communication. While the Port State Machine is a thorough extension of the Protocol State Machine in UML 2, our ServiceRoleStateMachine is just a very restricted Behavioral State Machine, constructed solely for use in our service modeling concept. The strength of the ServiceRoleStateMachine is its simplicity, which makes it easier to validate, while the Port State Machine is a versatile construction suitable for general modeling using UML.

Beyer et al. [2, 1] propose a formalism for web service interfaces based on automata, allowing to reason about the interfaces, checking for compatibility and substitutability. Our approach is not a complete formalism, but the behavioral contract as expressed by the ServiceRoleStateMachines is a formal representation of the visible interface behavior.

A theoretical framework for service and component modeling and composition is proposed by Broy et al. [4]), and like us they separate between service modeling and component modeling. They adopt the view of services as crosscutting interaction aspects of software systems, and thus put emphasis on interaction modeling.

An interaction-based service notion is also the basis for Ermagan and Krüger's UML profile for service modeling [6], which share our view that services are fundamentally defined by collaborations between components

6 Summary and Outlook

Our active service concept allows for peer-to-peer style communication, thereby not limiting service behavior to strictly one-way initiatives. We have presented an approach where we apply the active service concept into a SOA context by means of describing the services for publication and discovery. The active service notion builds upon some generalizations of what a SOA is, but does not breach any SOA principles.

It provides a duplex service concept with richer interfaces due to the two-way communication. Fundamental in our service concept is the behavioral contract expressing the expected and required behavior that the entities participating in the service must fulfill. Allowing services containing more than a single interface, and having to deal with the possibility of mixed initiatives may seem to increase the complexity of service design. In our view this is not the case. The complexiy comes from the nature of the services, not from the modeling.The modeling approach presented here enables us to explicitly address problems that are inherent in the service functionality.

In this paper we have focused on describing the functional properties of services for publication, using the UML Collaboration as a structural container for our behavioral contract. We are not excluding the other "usual" information published for discovery, such as keywords, semantic annotations [8] or service goals [22], and we will continue our work with exploring how to combine these into our approach. We have not elaborated on the notion of compatibility apart from stating that components are asserted to be validated against (and thereby fulfill) the published behavioral contracts.

References

1. Beyer, D., Chakrabarti, A., Henzinger, T.A.: An interface formalism for web services. In: Proceedings of the First International Workshop on Foundations of Interface Technologies (FIT 2005), San Francisco, CA, August 21 (2005)
2. Beyer, D., Chakrabarti, A., Henzinger, T.A.: Web service interfaces. In: WWW 2005: Proceedings of the 14th international conference on World Wide Web, pp. 148–159. ACM, New York (2005), `http://doi.acm.org/10.1145/1060745.1060770`
3. Box, D.: A guide to developing and running connected systems with indigo. MSDN Magazine 19(1) (January 2004), `http://msdn.microsoft.com/msdnmag/issues/04/01/Indigo/default.aspx`
4. Broy, M., Krüger, I.H., Meisinger, M.: A formal model of services. ACM Transactions Software Engineering and Methodology 16(1) (2007), `http://doi.acm.org/10.1145/1189748.1189753`
5. Erl, T.: Service-Oriented Architecture: Concepts, Technology, and Design. Prentice-Hall, Englewood Cliffs (2005)
6. Ermagan, V., Krüger, I.H.: A uml2 profile for service modeling. In: Engels, G., Opdyke, B., Schmidt, D.C., Weil, F. (eds.) MODELS 2007. LNCS, vol. 4735, pp. 360–374. Springer, Heidelberg (2007)
7. Fallside, D.C., Walmsley, P.: XML schema part 0: Primer second edition. W3C recommendation, W3C (2004), `http://www.w3.org/TR/2004/REC-xmlschema-0-20041028/`
8. Farrell, J., Lausen, H.: Semantic annotations for WSDL and XML schema. W3C recommendation, W3C (2007), `http://www.w3.org/TR/2007/REC-sawsdl-20070828/`
9. Floch, J.: Towards Plug and Play Services, Design and Validation of Roles. Ph.D. thesis, Department of Telematics, Faculty of Information Technology, Mathematics and Electrical Engineering, NTNU (2003)
10. Floch, J., Bræk, R.: ICT convergence: Modeling issues. In: Amyot, D., Williams, A.W. (eds.) SAM 2004. LNCS, vol. 3319, pp. 237–256. Springer, Heidelberg (2005)
11. Group, P.: Parlay X Web Services Specification, Version 2.1 - Short Messaging (2006), `http://www.parlay.org/en/specifications/pxws.asp`

12. Holzmann, G.J.: The SPIN model checker: Primer and reference manual. Addison-Wesley, Reading (2004)
13. Kraemer, F.A.: Arctis and Ramses: Tool Suites for Rapid Service Engineering. In: Proceedings of NIK 2007 (Norsk informatikkonferanse), Oslo, Norway, Tapir Akademisk Forlag (2007)
14. Kraemer, F.A., Herrmann, P., Bræk, R.: Aligning UML 2.0 state machines and temporal logic for the efficient execution of services. In: Meersman, R., Tari, Z. (eds.) OTM 2006. LNCS, vol. 4276, pp. 1613–1632. Springer, Heidelberg (2006)
15. Lewis, A.A.: Web services description language (WSDL) version 2.0: Additional MEPs. W3C note, W3C (2007), `http://www.w3.org/TR/2007/NOTE-wsdl20-additional-meps-20070626`
16. Liu, C.K., Booth, D.: Web services description language (WSDL) version 2.0 part 0: Primer. W3C recommendation, W3C (2007), `http://www.w3.org/TR/2007/REC-wsdl20-primer-20070626`
17. Mencl, V.: Specifying component behavior with port state machines. Electronic Notes in Theoretical Computer Science 101C, 129–153 (2004), `citeseer.ist.psu.edu/article/mencl04specifying.html`. In: de Boer, F., Bonsangue, M. (eds.): Special issue: Proceedings of the Workshop on the Compositional Verification of UML Models CVUML
18. Object Management Group: Unified Modeling Language 2.0 Superstructure Specification (2006)
19. Orchard, D., Weerawarana, S., Lewis, A.A., Moreau, J.J., Haas, H., Chinnici, R.: Web services description language (WSDL) version 2.0 part 2: Adjuncts. W3C recommendation, W3C (2007), `http://www.w3.org/TR/2007/REC-wsdl20-adjuncts-20070626`
20. Quartel, D.A.C., Steen, M.W.A., Pokraev, S., van Sinderen, M.: COSMO: A conceptual framework for service modelling and refinement. Information Systems Frontiers 9(2-3), 225–244 (2007), `http://dx.doi.org/10.1007/s10796-007-9034-7`
21. Samset, H., Bræk, R.: Dealing with Active and Stateful Services in Service-oriented Architecture. In: First International Workshop on Telecom Service Oriented Architectures (TSOA 2007). LNCS, Springer, Heidelberg (2008)
22. Sanders, R.T., Bræk, R., Bochmann, G., Amyot, D.: Service discovery and component reuse with semantic interfaces. In: Prinz, A., Reed, R., Reed, J. (eds.) SDL 2005. LNCS, vol. 3530. Springer, Heidelberg (2005)
23. Vissers, C.A., Logrippo, L.: The importance of the service concept in the design of data communications protocols. In: Diaz, M. (ed.) Protocol Specification, Testing and Verification V, Proceedings of the IFIP WG6.1 Fifth International Conference on Protocol Specification, Testing and Verification, Toulouse-Moissac, France, June 10-13, 1985, pp. 3–17. North-Holland, Amsterdam (1985)

Refactoring-Based Adaptation of Adaptation Specifications

Ilie Şavga and Michael Rudolf

Institut für Software- und Multimediatechologie,
Technische Universität Dresden, Germany
{is13,s0600108}@inf.tu-dresden.de

Summary. When a new release of a component replaces one of its older versions (*component upgrade*), changes to its interface may invalidate existing component-based applications and require adaptation. To automate the latter, developers usually need to provide *adaptation specifications.* Whereas writing such specifications is cumbersome and error-prone, maintaining them is even harder, because any evolutionary component change may invalidate existing specifications. We show how the use of a history of structural component changes (*refactorings*) enables automatic adaptation of existing adaptation specifications; the latter are written once and need not be maintained.

1 Introduction

A software framework is a software component that embodies a skeleton solution for a family of related software products [17]. To instantiate it to a concrete application, developers write custom-specific modules (*plugins*) that subclass and instantiate public types of the framework's Application Programming Interface (API). Frameworks are software artifacts, which evolve considerably due to new or changed requirements, bug fixing, or quality improvement. If changes affect their API, they may invalidate existing plugins; that is, plugin sources cannot be recompiled or plugin binaries cannot be linked and run with a new framework release. When upgrading to a new framework version, developers are forced to either manually adapt plugins or write update patches. Both tasks are usually error-prone and expensive, the costs often becoming unacceptable in case of large and complex frameworks.

To reduce the costs of component upgrade, a number of techniques propose to, at least partially, automate component adaptation [7, 9, 11, 16, 18, 21, 24]. These approaches rely on and require additional *adaptation specifications* that are used to intrusively update components' sources or binaries. Component developers have to manually provide such specifications either as different annotations within the component's source code [7, 11, 21] or in a separate specification [9, 16, 18, 24] used by a transformation engine to adapt the old application code. However, in case of large and complex legacy applications the cumbersome task of writing specifications is expensive and error-prone.

R. Lee (Ed.): Soft. Eng. Research, Management & Applications, SCI 150, pp. 189–203, 2008.
springerlink.com

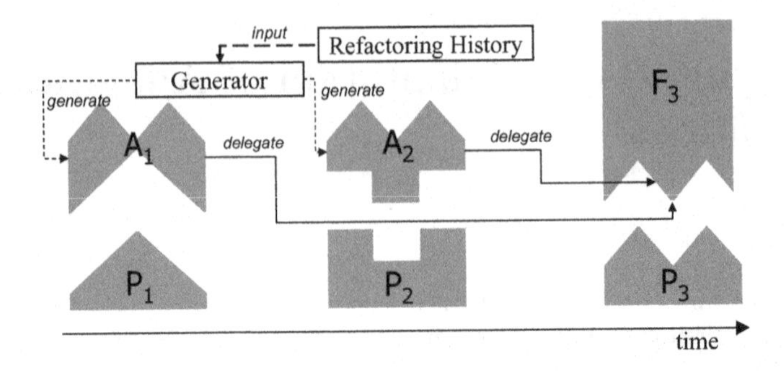

Fig. 1. Refactoring-based plugin adaptation. While a new plugin P_3 can be developed using the latest framework version F_3, existing plugins P_1 and P_2 are protected by adapters A_1 and A_2.

To alleviate the task of writing adaptation specifications, Dig and Johnson [13] suggest to reuse the information about the change to automatically perform adaptation. They investigated the evolution of five big software components, such as Eclipse [3], and discovered that most (from 81% up to 97%) problem-causing API changes were *refactorings* – behavior-preserving source transformations [19]. Common examples of refactorings include renaming classes and members to better reflect their meaning, moving members to decrease coupling and increase cohesion, and adding new and removing unused members. As at the time of framework refactoring its plugins are usually not available for update, they are invalidated by refactoring.

Based on the results of [13], two techniques proposed independently by Dig et al. in [14] and by us in [22] use refactoring information to *insulate* plugins from framework changes. The refactoring information required can be logged automatically (e.g., in Eclipse [3] or JBuilder [1]) or harvested semi-automatically [12] and thus does not require additional specifications from developers. Figure 1 shows, how we use the refactoring history to create adapters between a framework and some of its plugins upon the release of a new framework version. The adapters then shield the plugins by representing the public types of the old framework version, while delegating to the latest one.

Although in general the refactoring-based adaptation is capable of adapting more than 80% of problem-causing API changes, its restriction to pure structural changes also entails its limitations. Other changes affecting existing plugins require either manual or specification-based adaptation. For example, in a new framework release developers may change the way message exchanges between the framework and its plugins are performed. They introduce a new framework method and require calling it (possibly, with a default parameter) from plugins. This change leads to a *protocol mismatch* characterized by improper message exchange among components [8], in this case between the new framework and its old plugins that do not call the new method. To bridge protocol mismatches, developers have to specify messages that components may send and receive as well as the valid message exchange. Basing on these specifications *protocol adapters* that support proper intercomponent communication are generated (e.g., [20, 24]).

However, once such specifications are provided, component evolution unavoidably demands their *maintenance*, because evolutionary changes may alter component parts on which existing specifications rely, rendering the latter useless. For instance, if a framework type that is referred to in a specification is renamed, the specification is no longer valid. In such cases, specifications must be updated along with the change of the involved components, which raises the complexity and costs of component adaptation.

Extending our refactoring-based approach [22], we want to maximally ease the task of adapting remaining changes. For protocol changes, the corresponding specifications must be undemanding to write and not require maintenance throughout subsequent component evolution. The main contribution of this paper is in enabling component developers to write protocol adaptation specifications that are:

- in-time: specified at the actual time of component change. It is inherently easier to specify the adaptation of small incremental changes upon their application than to write large specifications involving complex component and change dependencies in one big step upon a new component release.
- durable: valid at the time of their execution regardless of any component change applied after the specification was written. This eliminates the need to maintain specifications.

Given a protocol adaptation specification, we use the information about subsequent structural component changes to perform the aforementioned refactoring-based adaptation and to shield *the specification* from those changes. In this way, we adapt existing specifications in the sense that we preserve their validity in the context of component evolution.

In Sect. 2 we sketch the refactoring-based adaptation by describing its main concepts used throughout the rest of the paper. Section 3 discusses the notion of protocol adapters in detail and continues with our main contribution – adaptation of existing adaptation specifications using refactoring information – followed by the discussion of relevant issues and empirical results in Sect. 4. We overview related work in Sect. 5 and conclude in Sect. 6.

2 Refactoring-Based Adaptation

Although refactoring preserves behavior, it changes syntactic component representation, on which other components may rely. Figure 2 shows the application of the `ExtractSubclass` refactoring to a component class modeling network nodes.[1] If an existing component *LAN* calling the method *broadcast* on an instance of *Node* is not available for update, it will fail to both recompile and link

[1] This example was inspired by the LAN simulation lab used at the University of Antwerp for teaching refactoring [5]. For simplicity we omit several node methods (e.g., *send* and *receive*).

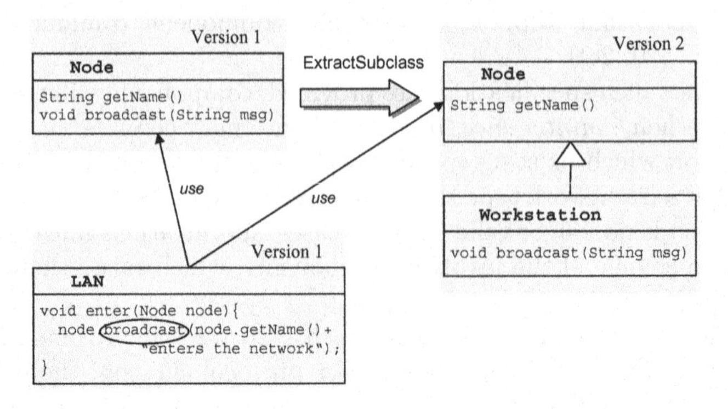

Fig. 2. `ExtractSubclass` refactoring. The method *broadcast* is moved from *Node* to its new subclass *Workstation* and cannot be located from the existing *LAN*.

with the new *Node* version. As a consequence, any application using these two components will be broken by the upgrade of *Node*.

2.1 Comebacks

In [22] we formally define our refactoring-based adaptation, so that the adapters for the API refactorings could be constructed automatically and the soundness of the adaptation could be ensured. Effectively, we roll back the changes introduced by framework refactorings by executing their inverses. We cannot inverse directly on framework types, because we want new plugins to use the improved framework. Instead, we create adapters (one for each framework API type) and then inverse refactorings on adapters. We call these inverses *comebacks*. For our running example of `ExtractSubclass`, the compensating (object) adapter constructed by the comeback of the refactoring is shown in Fig. 3.

Technically, a comeback is realized in terms of refactoring operators executed on adapters. It is defined as a template solution and instantiated to an executable specification by reusing parameters of the corresponding refactoring. For some refactorings, the corresponding comebacks are simple and implemented by a single refactoring. For example, to the refactoring `RenameClass(name, newName)` corresponds a comeback consisting of a refactoring `RenameClass(newName, name)`, which renames the adapter to the old name. For other refactorings, their comebacks consist of sequences of refactorings. For instance, the comeback of `PushDownMethod` is defined by the `DeleteMethod` and `AddMethod` refactoring operators, the sequential execution of which effectively moves (pushes up) the method between adapters. Moreover, complex comebacks may be defined by composing other, more primitives comebacks. This is the case for the comeback of `ExtractSubclass`, which is defined by combining the comebacks of `PushDownMethod` and `AddClass`.

For an ordered set of refactorings that occurred between two component versions, the execution of the corresponding comebacks in reverse order yields the

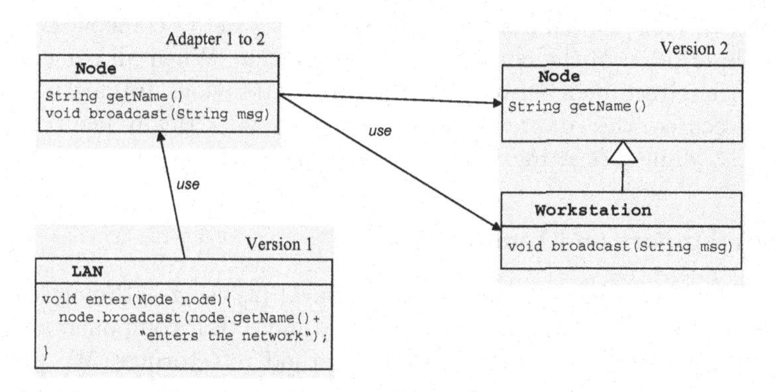

Fig. 3. Compensating adapter. The adapter represents the old *Node* and delegates to the appropriate component methods.

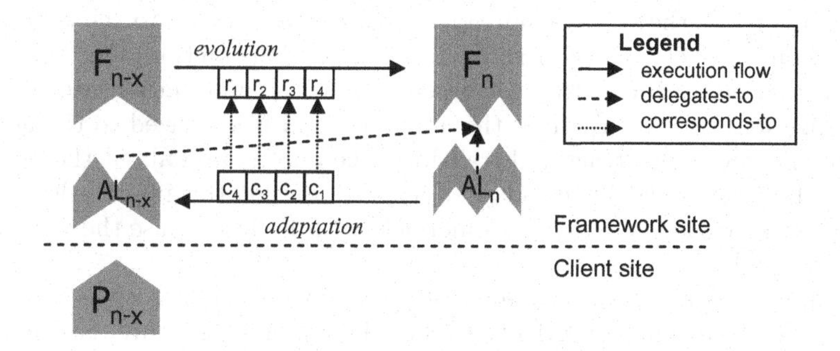

Fig. 4. Adaptation workflow. To a set of refactorings (r_1–r_4) between two framework versions correspond comebacks (c_4–c_1). Comebacks are executed on the adaptation layer AL_n backwards to the framework refactorings. The resulting adaptation layer AL_{n-x} delegates to the latest framework version, while adapting plugins of version P_{n-x}.

adaptation layer. Figure 4 shows the workflow of refactoring-based adaptation in case the component being upgraded is a framework.[2] In a nutshell, we copy the latest API of the framework and inverse all refactorings on it, so that the old API (mimicked by the adaptation layer) is reconstructed fully automatically. First, we create the adaptation layer AL_n (the right part of the figure) that is a full copy of the latest framework API delegating to the latest framework version. Therefore for each API class of the latest framework version F_n we provide an adapter with exactly the same name and set of method signatures. An adapter delegates to its public class, which becomes the adapter's delegatee. Once the adapters are created, the actual adaptation is performed by executing comebacks backwards

[2] Although we focus on object-oriented frameworks, the technique is similarly applicable to the upgrade of other types of object-oriented components, such as software libraries.

with respect to the recorded framework refactorings, where a comeback is derived using the description of the corresponding refactoring. When all comebacks for the refactorings recorded between the last and a previous framework version F_{n-x} are executed, the adaptation layer AL_{n-x} is syntactically identical to the API of F_{n-x}, while delegating to the newest framework version.

2.2 Tool Support and Limitations

We implemented our adaptation tool ComeBack! [2] using a Prolog logic programming engine. Currently we support Java and .NET components and can compensate for twelve common class and method refactorings. We provide a comeback library consisting of the corresponding comeback transformations specified as Prolog rules. Given the latest framework binaries, the information about the API types (type and method names, method signatures, inheritance relations) is parsed into a Prolog fact base. After examining the history of framework refactorings, the corresponding comebacks are loaded into the engine and executed on the fact base as described in the previous section. Once all comebacks have been executed, the fact base contains the information necessary for generating adapters (it describes the adapters) and is serialized to the adapter binaries. Thereby we extract and transform the information about the program and not the program itself; the adapter generation using this information is then the final step. We also combined ComeBack! with Eclipse to use the refactoring history of the latter.

Because comebacks are executed on adapters, for some refactorings comebacks cannot be defined due to particularities of the adapter structure. For instance, it is not possible to define a comeback for field refactorings (e.g., renaming and moving public fields), because adapters cannot adapt fields directly. Instead, we require API fields to be encapsulated. Moreover, for certain refactorings (e.g., adding parameters), developers are prompted for additional information (e.g., default parameter value) used to parameterize comebacks. In [23] we are making developers aware of such limitations and suggest solutions whenever possible.

3 Adapting Protocol Adaptation Specifications

We envisage the combination of the refactoring-based adaptation with other adaptation techniques. In this section we discuss, how several important problems associated with *protocol adaptation* are solved by its integration with the comeback approach. We start with a short introduction to and a running example of protocol adaptation, discuss then problems introduced by subsequent component refactorings, and stipulate a refactoring-based solution. Since our intention is to show problems and adopted solutions associated with the creation and maintenance of protocol adaptation specifications, and not the specifications themselves, we use a simplified example without going too much into the details of the specification formalism used.

3.1 Protocol Adaptation

Modifications of a component may affect its behavior, that is, the way it interoperates with other components (by exchange of method calls), leading to protocol mismatches. Common examples of such mismatches are [8]:

- Non-matching message ordering. Although components exchange the same kind of messages, their sequences are permuted.
- Surplus of messages. A component sends a message that is neither expected by the connected component nor necessary to fulfill the purpose of the interaction.
- Absence of messages. A component requires additional messages to fulfill the purpose of the interaction. The message content can be determined from outside.

Consider the two components of Fig. 5.1 (their APIs are not shown in the figure). The *LAN* component models a Token Ring local area network that accepts and removes nodes via the methods *enter(Node node)* and *exit(Node node)* and asks a node to create and forward a new packet using the method *broadcast(String msg)*. The *Node* component abstracts a separate network node and provides, among others, the methods to enter and exit the network (*enter(Node thisNode)* and *exit(Node thisNode)*). In addition, it has a method *broadcast(String msg)* that creates a packet from the given message, marks the packet with the ID of the creator node, forwards it to the successor node, and acknowledges (*broadcastAck()*) back to the network. Since both components are of the same release 1, their APIs and protocols match, so that there is no need for adaptation.

In the next release the developers generalized *Node* to increase its reusability for other types of networks. In particular, they introduced a two-phase protocol for entering the network: the node first asks for a permission to enter a network *enterReq(Node thisNode)*, passing in itself as the argument, and only if allowed actually enters the network. The same changes were applied to the protocol for

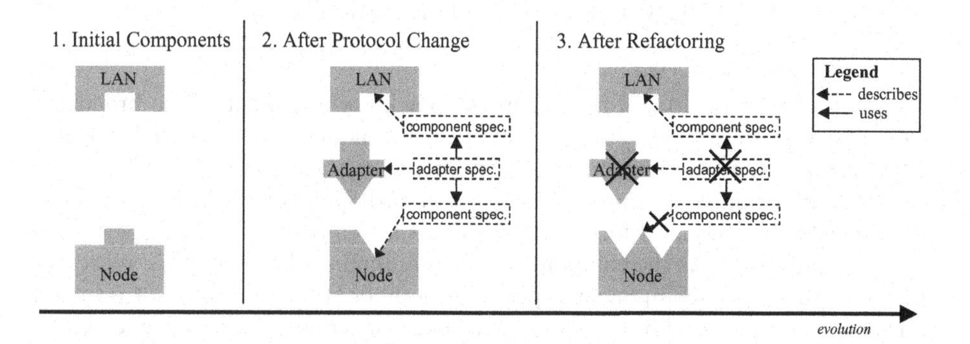

Fig. 5. Component Evolution. Initially the components cooperate as intended. The second release of *Node* requires adaptation specifications, which are then invalidated by a subsequent refactoring.

```
1  Collaboration Node {
2    Receive Messages{
3      mayEnter();
4      maynotEnter(String:why);
5      mayExit();
6      maynotExit(String:why);
7      broadcast(String message);
8    };
9    Send Messages{
10     enterReq(Node thisNode);
11     enter();
12     exitReq(Node thisNode);
13     exit();
14     ceased(Packet packet);
15   };
16   Protocol{
17     States{1(init),2,3,4,5};
18     1: -enterReq -> 2;
19     2: +maynotEnter -> 1;
20     2: +mayEnter -> 3;
21     3: +broadcast -> 3;
22     3: -ceased -> 3;
23     3: -exitReq -> 4
24     4: +mayExit -> 6;
25     5: +maynotExit -> 3;
26     6: -exit -> 1;
27   }
28 } Collaboration LAN {
29   Receive Messages{
30     enter(Node node);
31     exit(Node node);
32     broadcastAck();
33   };
34   Send Messages{
35     broadcast(String message);
36   };
37   Protocol{
38     States{A(init),B};
39     A: +enter -> A;
40     A: +exit -> A;
41     A: -broadcast -> B;
42     B: +broadcastAck -> A;
43   }
44 }
```

List. 1. Component Behavior Specification

leaving the network. In addition, there is no acknowledgement of the broadcast call sent back from *Node* anymore. Instead, the node issues a *ceased* message when it removes its previously created packet from the network.

This is an example of a protocol mismatch: although the components possess the functionality required for an interaction, they cannot interact properly and require adaptation. As the component API is limited to the syntactical representation of the public component types and methods, the adaptation of protocol mismatches requires additional specification of the components' behavior interface and of the valid mapping of component messages (Fig. 5.2). Listing 1 shows the behavior interfaces of *LAN* and *Node* specified using the (simplified) formalism of Yellin and Strom [24]. The interfaces (called *collaborations* in the listing)

```
1  /*LEGEND M0={} M1={thisNode} M2={message}*/
2  <1,A,M0>: +enterReq from Node -> <2,A,M1>, write(thisNode);
3  <2,A,M1>: -mayEnter to Node -> <2,A,M1>;
4  <2,A,M1>: +enter from Node -> <3,A,M1>;
5  <3,A,M1>: -enter to LAN -> <3,A,M0>,
6            node = read(thisNode), invalidate(thisNode);
7  <3,A,M0>: +broadcast from LAN -> <3,A,M2>, write(message);
8  <3,A,M2>: -broadcast to Node -> <3,B,M0>,
9            message = read(message), invalidate(message);
10 <3,B,M0>: +broadcastAck to LAN -> <3,A,M0>;
11 <3,A,M0>: -ceased from Node -> <3,A,M0>;
12 <3,A,M0>: -exitReq from Node -> <4,A,M1>, write(thisNode);
13 <4,A,M1>: +mayExit to Node -> <4,A,M1>;
14 <4,A,M1>: -exit from Node -> <5,A,M1>;
15 <5,A,M1>: +exit to LAN -> <5,A,M0>,
16           node = read(thisNode), invalidate(thisNode);
```

List. 2. Mapping Specification

are described as a set of sent and received messages augmented by a finite state machine specification. The latter defines the legal sequences of messages that can be exchanged between a component and its mate (*init* stands for the initial state, - for sending and + for receiving messages). For instance, after sending the enter request, *Node* may only accept either *maynotEnter* (line 19) or *mayEnter* (line 20) messages and, in the latter case, can enter the network and broadcast packets (line 21). For clarity, the states of *Node* are marked by numbers (1–6) and those of *LAN* by letters (A, B).

For these two behavior interfaces List. 2 provides a mapping specification *NodeLAN* relating their states and messages. For each permitted combination of component states (numbers of *Node* and letters of *LAN* enclosed in angle brackets) it specifies the allowed transitions of both components and the valid memory state M of messages being exchanged. For instance, after receiving an enter request from a node (line 2), the adapter by default allows the node to enter the network (line 4) and, when getting the node's *enter* message, actually injects it into the LAN (line 6). To keep track of the components' states, the adapter updates them correspondingly (again enclosed in angle brackets). To properly exchange the message data, certain method parameters need to be saved by the adapter. For example, *thisNode* sent by *Node.enterReq* may be sent to *LAN* only when the node actually enters the network (execution of *Node.enter()*). Meanwhile the adapter saves the parameter internally (shown as a call *write* to the implicitly generalized virtual memory). When the saved parameters are not needed anymore, they are deleted from the virtual memory by the adapter, as denoted by *invalidate.* Note also the handling of message absence (*broadcastAck* to *LAN*, *mayEnter* and *mayExit* to *Node*) and surplus (*ceased* from *Node*).

3.2 Refactorings Invalidate Specifications

Assume the specification is valid and the second release of the component was deployed successfully. However, in release 3 the developers realized that the *broadcast* method does not belong to every node of a network, but only to certain ones. In particular, workstation nodes may create and then broadcast packets, but print servers may not. To properly model this design requirement and avoid the abuse of the method, the developers applied the `ExtractSubclass` refactoring, effectively subclassing *Node* with a *Workstation* class and moving (pushing down) the method *broadcast* to the latter, as shown in Fig. 2 on page 4.

The refactoring will clearly invalidate the interface specification of *Node* (lines 7 and 21) and, hence, a part (lines 7–9) of the existing *NodeLAN* specification (Fig. 5.3). The situation aggravates with the growing number of releases and chaining of interdependent modifications. For instance, the developers of *Node* could later (1) rename (`RenameMethod` refactoring) the method *broadcast* to *originate* in order to reuse it also for multicasting, and (2) add a parameter to *maynotEnter* (`AddParameter` refactoring) to specify the reason of rejecting the entrance.

In general, any syntactic change affecting components referred to in specifications will invalidate the latter. The reason is that by changing the syntactical representation, refactoring modifies the initial context on which the specifications depend. Either the specifications have to be updated to match the new context or the context itself has to be recovered. We perform the latter using comebacks.

3.3 Rescuing Specifications by Comebacks

As described in Sect. 2, the execution of comebacks in reverse order with regard to their corresponding refactorings yields an adaptation layer, which wraps new component functionality in terms of an old API syntax. Besides automatically bridging the signature mismatches, this execution leads to another important result – it reestablishes the context for previously written adaptation specifications. The intuition is that by effectively inverting refactorings on adapters, comebacks reconstruct the "right" syntactic names used in the specifications [22].

Figure 6.1 shows the state of adaptation before executing the comeback of `ExtractSubclass`. In between the *LAN* and *Node* components there is the adapter component (also called *Node*), possibly created in a step-wise manner by previously executed comebacks. At this stage, the *NodeLAN* specification is not (yet) valid. Now the comeback of `ExtractSubclass` transforms the adapter and leads to the situation shown in Fig. 6.2. Because the comeback inverted its corresponding refactoring on the adapter, the latter has exactly the same syntactic form as the original component *Node* before `ExtractSubclass` in Fig. 5.2. At this stage *NodeLAN* becomes valid again and can be used to derive the corresponding adapter. Intuitively, the comebacks reconstruct backwards the public types to which the specification relates; the only difference is that now (some of) these types are adapters. As a consequence, the developer can specify durable adapters at the time of the actual change.

There is still an important difference between the two middle parts of the Figs. 5 and 6: while in Fig. 5.2 the adapter is generated between two initial components, in Fig. 6.2 it is generated between the LAN component and the

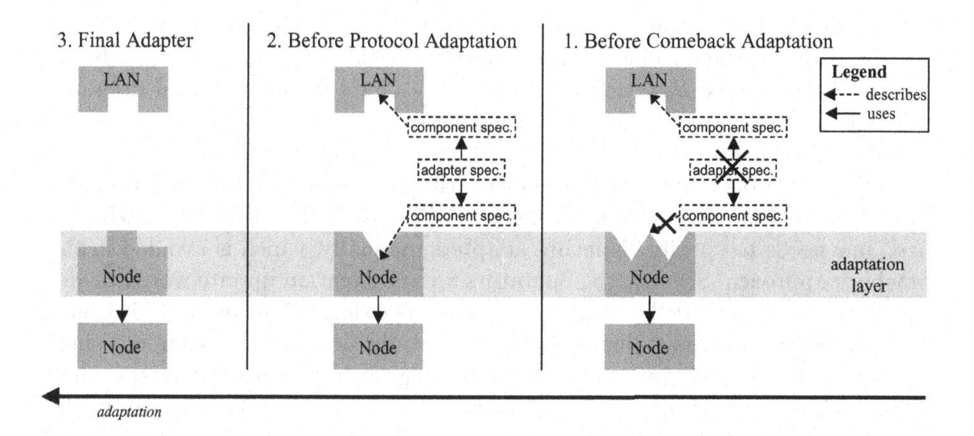

Fig. 6. Component Adaptation. The comeback execution reestablishes the context of adaptation specifications. They can then be executed directly on adapters to avoid performance overhead.

previously derived adapter leading to stacking of adapters. Although not affecting the actual functionality, adapter stacking may considerably decrease the performance due to the additional layer of redirection. In systems where the quality of service requires a certain level of performance, the performance penalties may become unacceptable.

In our ComeBack! tool we address this performance issue by integrating the protocol adaptation machinery into the existing refactoring-based environment. Given protocol adaptation specifications, the validity of which is proved at the specification level, we translate them into a corresponding set of Prolog facts and rules. The execution of these rules at the time of protocol adaptation updates the (adapter) fact base created by previously executed comebacks and protocol adapters. Because the description of the protocol adapter is thus effectively embedded into the fact base, no separate adapter is required (Fig. 6.3).[3]

Technically, because we assume correctness of specifications, we do not have to implement their state machines. By construction, the client components behave according to the state automaton before the upgrade and a valid specification does not change that. In fact, the only state the generated adapters maintain at runtime is the pointer to their delegatees and a storage for temporarily saved messages to implement the *write* and *read* operations of List. 2. As a consequence, the task of integrating both adaptation approaches is reduced to generating custom adapter methods that (1) read and write message contents to a local backing-store, such as a hashtable (to save and retrieve data exchanged), (2) perform actual delegation, (3) send default messages (in case of message absence), or (4) do nothing (in case of message surplus).

[3] One of the main requirements of our technology – the total order of changes in the component history – is realized using timestamps.

As an alternative to the backward comeback execution that recovers the specification context, the specifications themselves could be updated to match the updated components. For each refactoring one could define a special operator to update affected specifications. A (forward) execution of these operators along the refactoring history would produce the correct specifications. Although this approach is appropriate for the formalisms that combine signature and protocol adaptation in the same specifications (e.g., [9, 16]), it has several drawbacks. First, one needs to specify signature adapters manually, which is avoided in the comeback approach. Second, the operators for specification update would be dependent on and need to be rewritten for any new formalism used. Third, and most important, the formalisms themselves would need to be extended, since none of them provides any means for reflecting such changes in the specifications (e.g., that some functionality of *Node* is to be found in *Workstation* in the new release).

4 Evaluation

We performed two case studies using small-to-medium size frameworks. Using API documentation and clients, for each framework we investigated changes between four of its major versions. Gradually, the plugins of the first major version were manually adapted to compile and run with the second, third and fourth framework versions. Whenever possible, such adaptation was performed by refactoring. In such a way we discovered exactly the backward-incompatible framework changes and, where possible, modeled them as refactorings. For all refactorings detected the corresponding comebacks were specified. For the remaining changes we investigated whether they affected message exchange between clients and the framework.

In the first case study we investigated JHotDraw [4], a well-known Java GUI framework. We used its versions 5.2, 5.3, 5.4, and 6.0 as well as four sample clients delivered with the version 5.2. We discovered eight backward-incompatible changes between versions 5.2 and 5.3, three changes between 5.3 and 5.4 and one change between 5.4 and 6.0. In total 12 changes were discovered, 11 (92%) of them being refactorings. The exception was the change of a collection type from the built-in Java *Enumerator* to the JHotDraw user-defined *FigureEnumerator*. It could neither be modeled as a refactoring nor as a protocol mismatch.

For the second case study we used SalesPoint [6], a framework developed and used at our department for teaching framework technologies. We used the framework versions 1.0, 2.0, 3.0 and 3.1 as well as two clients (student exercises) developed with the framework version 1.0. The first client reused mostly business logic of the framework. In total 48 changes were discovered, 47 (97.9%) of them being refactorings. The remaining one was the change of a default action (throwing an exception instead of doing nothing) and we compensated for it with the help of the corresponding protocol adapter.

Besides reusing the framework's business logic, the second SalesPoint client also made heavy use of the framework's GUI facilities, which changed considerably between the framework versions. Whereas in version 1.0 the GUI event handling was implemented solely by the framework, in version 2.0 (due to the switch to a new Java version) its implementation was based on the Java event handling. As a consequence, for the second client only about 70% (49 in total) of the changes could be considered refactorings. About 90% (16 in total) of the remaining backward-incompatible changes were protocol changes, in particular, of event handling in the GUI. Since we intended to explore the possibility of combining our refactoring-based adaptation with protocol adaptation, and not the full-fledged implementation of the latter, we implemented protocol adapters only for two selected protocol mismatches.

All in all, our results confirm the importance of refactorings for API evolution of software components previously pointed out by Dig and Johnson [13]. Moreover, in our case studies most of the remaining changes beyond refactorings could be modeled as protocol changes. However, the complexity of protocol adaptation in combination with refactoring-based adaptation may vary considerably depending on change particularities and therefore needs further investigation.

5 Related Work

As mentioned in the Introduction, most of the adaptation approaches [7, 9, 11, 16, 18, 21, 24] require additional adaptation specifications from developers. Recent research focused on how specific properties of certain modifications (e.g., behavior preservation of refactorings, deadlock-free protocol changes) can be proved statically and used to automate adaptation. For instance, the "catch-and-replay" approach [15] records the refactorings applied in an IDE as a log file. The file can be either delivered to the application developer, who "replays" refactorings on the application code (invasive adaptation), or used to generate binary adapters [22, 14]. These approaches, however, do not discuss adaptation of changes beyond refactorings.

In the same way, behavior adaptation is driven by the idea of capturing important behavioral component properties to reason about protocol compatibility, deadlock freedom, and synchronization of distributed components as well as to (semi-) automatically derive adapters and check for their correctness. Building on the seminal paper of [24], a number of approaches used, for example, label transition systems [10], message system charts augmented with temporal logic specifications [16], or process algebra [9] for specification of behavior and adapters. Although several approaches (most notably [9, 16]) address signature mismatches as well, their specifications are embedded into behavior specifications and suffer from the same maintenance problems we discussed in this paper.

6 Conclusion

Besides automatically adapting most of the component changes, our comeback approach fosters component maintenance in two ways. First, because refactoring-based adaptation only requires the new component API and the refactoring history, developers do not need to specify and later maintain specifications to compensate for structural changes. Second, for remaining changes of components' behavior developers are able to write in-time and durable protocol adaptation specifications, which are easier to specify and do not need to be maintained. Their automatic adaptation is implied by the particularities of our approach and comes "for free," as no additional means is required. We stipulate that the so-alleviated adaptation and maintenance not only reduce the costs and improve the quality of component upgrade but also relax the constraints on the permitted API changes allowing for appropriate component evolution.

Acknowledgement. We thank Mirko Seifert and our anonymous SERA'08 reviewers for their valuable comments.

References

1. Borland JBuilder, `http://www.codegear.com/products/jbuilder`
2. ComeBack! homepage, `http://comeback.sf.net`
3. Eclipse Foundation, `http://www.eclipse.org`
4. JHotDraw framework, `http://www.jhotdraw.org`
5. LAN-simulation lab session, `http://www.lore.ua.ac.be`
6. SalesPoint homepage, `http://www-st.inf.tu-dresden.de/SalesPoint/v3.1`
7. Balaban, I., Tip, F., Fuhrer, R.: Refactoring support for class library migration. In: OOPSLA 2005: Proceedings of the 20th annual ACM SIGPLAN conference on Object oriented programming, systems, languages, and applications, pp. 265–279. ACM Press, New York (2005)
8. Becker, S., Brogi, A., Gorton, I., Overhage, S., Romanovsky, A., Tivoli, M.: Towards an engineering approach to component adaptation. In: Reussner, R., Stafford, J.A., Szyperski, C.A. (eds.) Architecting Systems with Trustworthy Components. LNCS, vol. 3938, pp. 193–215. Springer, Heidelberg (2006)
9. Brogi, A., Canal, C., Pimentel, E.: Component adaptation through flexible subservicing. Science of Computer Programming 63(1), 39–56 (2006)
10. Canal, C., Poizat, P., Salaün, G.: Synchronizing behavioural mismatch in software composition. In: Gorrieri, R., Wehrheim, H. (eds.) FMOODS 2006. LNCS, vol. 4037, pp. 63–77. Springer, Heidelberg (2006)
11. Chow, K., Notkin, D.: Semi-automatic update of applications in response to library changes. In: ICSM 1996: Proceedings of the 1996 International Conference on Software Maintenance, p. 359. IEEE Computer Society Press, Washington (1996)
12. Dig, D., Comertoglu, C., Marinov, D., Johnson, R.: Automated detection of refactorings in evolving components. In: Thomas, D. (ed.) ECOOP 2006. LNCS, vol. 4067, pp. 404–428. Springer, Heidelberg (2006)
13. Dig, D., Johnson, R.: The role of refactorings in API evolution. In: ICSM 2005: Proceedings of the 21st IEEE International Conference on Software Maintenance (ICSM 2005), pp. 389–398. IEEE Computer Society Press, Washington (2005)
14. Dig, D., Negara, S., Mohindra, V., Johnson, R.: ReBA: Refactoring-aware binary adaptation of evolving libraries. In: ICSE 2008: International Conference on Software Engineering. Leipzig, Germany, (to appear, 2008)
15. Henkel, J., Diwan, A.: Catchup!: capturing and replaying refactorings to support API evolution. In: ICSE 2005: Proceedings of the 27th International Conference on Software Engineering, pp. 274–283. ACM Press, New York (2005)

16. Inverardi, P., Tivoli, M.: Software architecture for correct components assembly. In: Bernardo, M., Inverardi, P. (eds.) SFM 2003. LNCS, vol. 2804, pp. 92–121. Springer, Heidelberg (2003)
17. Johnson, R., Foote, B.: Designing reusable classes. Journal of Object-Oriented Programming 1(2), 22–35 (1988)
18. Keller, R., Hölzle, U.: Binary component adaptation. In: Jul, E. (ed.) ECOOP 1998. LNCS, vol. 1445, pp. 307–329. Springer, Heidelberg (1998)
19. Opdyke, W.F.: Refactoring object-oriented frameworks. Ph.D. thesis, University of Illinois at Urbana-Champaign, Urbana-Champaign, IL, USA (1992)
20. Reussner, R.: The use of parameterised contracts for architecting systems with software components. In: Bosch, J., Weck, W., Szyperski, C. (eds.) WCOP 2001: Proceedings of the Sixth International Workshop on Component-Oriented Programming (2001)
21. Roock, S., Havenstein, A.: Refactoring tags for automatic refactoring of framework dependent applications. In: Wells, D., Williams, L. (eds.) XP 2002. LNCS, vol. 2418, pp. 182–185. Springer, Heidelberg (2002)
22. Şavga, I., Rudolf, M.: Refactoring-based adaptation for binary compatiblity in evolving frameworks. In: GPCE 2007: Proceedings of the Sixth International Conference on Generative Programming and Component Engineering, pp. 175–184. ACM Press, Salzburg (2007)
23. Şavga, I., Rudolf, M., Lehmann, J.: Controlled adaptation-oriented evolution of object-oriented components. In: IASTED SE 2008: Proceedings of the IASTED International Conference on Software Engineering. ACTA Press, Innsbruck (2008)
24. Yellin, D.M., Strom, R.E.: Protocol specifications and component adaptors. ACM TOPLAS: ACM Transactions on Programming Languages and Systems 19(2), 292–333 (1997)

Resolving Complexity and Interdependence in Software Project Management Antipatterns Using the Dependency Structure Matrix

Dimitrios Settas and Ioannis Stamelos

Dept. of Informatics
Aristotle University of Thessaloniki, Thessaloniki, Greece
dsettas@csd.auth.gr

Summary. Software project management antipatterns are usually related to other antipatterns and rarely appear in isolation. This fact introduces inevitable interdependence and complexity that can not be addressed using existing formalisms. To reduce this complexity and interdependence, this paper proposes the Dependency Structure Matrix (DSM) as a method that visualizes and analyzes the dependencies between related attributes of software project management antipatterns. Furthermore, DSM provides a methodology that can be used to visualize three different configurations that characterize antipatterns and resolve cyclic dependencies that are formed between interdependent antipattern attributes. The proposed framework can be used by software project managers in order to resolve antipatterns that occur in a software project in a timely manner. The approach is exemplified through a DSM of 25 attributes of 16 related software project management antipatterns that appear in the literature and the Web.

Keywords: Software Project Management Antipatterns, DSM.

1 Introduction

Software project management antipatterns are mechanisms that can greatly benefit software project managers by capturing and representing software project management knowledge but also by catering to its collaborative exchange. These mechanisms suggest commonly occurring solutions [1] to problems regarding dysfunctional behavior of managers or pervasive management practices that inhibit software project success [2]. Software project management antipatterns can manage all aspects of a software project more effectively by bringing insight into the causes, symptoms, consequences, and by providing successful repeatable solutions [3]. In this paper, we consider that software project management antipatterns can be related through the following attributes: causes, symptoms and consequences. The readers that are not familiar with antipatterns can use [2], [3] for an introduction to the topic.

Though it is widely accepted that software project management antipatterns rarely appear isolated [4] and can be related with other antipatterns [1], the issue of addressing iteration and analyzing the dependencies between related

R. Lee (Ed.): Soft. Eng. Research, Management & Applications, SCI 150, pp. 205–217, 2008.
springerlink.com

antipatterns has not been addressed. Mathematical [5] and semantic Web [6] representations that have been used to model software project management antipatterns, do not support this task because they can not be used in order to address the issue of visualizing, analyzing and resolving interdependent antipattern attribute relationships. Dependency Structure Matrix or Design Structure matrix (DSM) [7] is a system modeling tool that can represent system elements and their relationships in a compact way that highlights important patterns in the data (i.e feedback loops or clusters) [8].

DSM can be used to provide software project managers with a compact, matrix representation of a set of related antipattern attributes that exist during a software project. In this paper, DSM partitioning analysis is carried out in order to represent antipattern attribute relationships using three different DSM configurations. Partitioning is used to eliminate or reduce feedback marks that exist between antipattern attributes, thus reducing interdependence between related antipattern attributes. This analysis also provides a methodology for resolving interdependent relationships called "circuits" using assumptions [9].

The resulting DSM can be used to assist software project managers in addressing antipattern attributes in a specific sequence that reduces interdependence between antipattern attributes. By recognizing which other antipattern attributes are reliant upon information outputs generated by each attribute, DSM can illustrate the constituent attributes of software project management antipatterns and the corresponding dependencies. Hence, software project managers will be able to resolve antipatterns that occur in a software project in a timely manner.

There are numerous commercial tools that enable users to build DSM models and run the program calculations. For the purposes of the DSM models, i.e. for their graphical representations and various calculations, PSM32[1] software tool was employed. This paper is divided in 5 sections, which are organized as follows: section 2 describes the background, the related work and the literature review used in our research. Section 3 describes data collection and the DSM analysis used in this paper. Section 4 describes the results of the antipattern DSM and the interpretation of the data. Finally, in section 5, the findings are summarized, conclusions are drawn and future work is proposed.

2 Background and Related Work

2.1 Background

DSM is a popular representation and analysis tool for system modelling, especially for product architecture decomposition and integration [8]. A detailed description of DSM methodology is outside the scope of this paper. The readers that are not familiar with DSM can use [8], [9] for an introduction to the topic. DSMs are usually referred to as static or time-based. Static DSMs can either be component-based or people-based, while time-based DSMs can either

[1] http://www.problematics.com

be activity-based or parameter-based [8]. The main difference between static and time-based DSMs is that in static DSMs the relationships between elements are not time based. Furthermore, static based DSMs are analyzed using clustering, while time-based relationships are analyzed using partitioning algorithms that re-sequence the elements in the matrix, making it a lower triangular matrix [10]. This analysis is referred to as sequencing and is used in this paper in order to uncover the dependencies that define the truly coupled antipattern attributes.

Considering, the fact that the set of data used in this paper does not take time into account, component based DSMs are used in this paper. By considering antipatterns as products and antipattern attributes as components, the interactions between antipattern attributes are documented using DSM. In a component-based DSM the off-diagonal elements are clustered by reordering the rows and columns of the DSM in order to maximize the density of clusters of components along the diagonal. The objective is to maximize interactions between elements within clusters while minimizing interactions between clusters [8]. These clusters can integrate components (i.e. antipattern attributes) into "chunks". Clustering small and/or sparse matrices can be carried out by manually manipulating rows and columns of the matrix [8]. However, this approach is generally recommended for product development and hardware products because of the highly symmetric nature of many spatial and structural design dependencies between physical components [11]. Furthermore, in the case of antipatterns, clustering does not benefit antipatterns because software project managers are already aware of the attributes that belong to each antipattern when creating a DSM of antipattern attributes. Therefore, alternative DSM analysis needs to be carried out, which is described in the following section.

2.2 Related Work

A mapping usually exists between static and time-based DSMs [12]. System components can be regarded as distinct line items in a work breakdown structure and can therefore be regarded as separate design activities. In turn, these design activities can be assigned to separate teams. Furthermore, the relationship between a component-based and a parameter-based DSM is usually at the level of aggregation in describing a system [12]. In the case a component being fully characterized by a single parameter, then the component-DSM and parameter-DSM for such a representation would coincide. Therefore, both static and time-based DSM analysis can be used in cases in which a component is characterized by a single parameter. This is the case in this paper and for this reason, time based analysis (partitioning) in a component-based DSM has been applied.

DSM methodology has been used in the software domain in order to capture the interactions between class functions used by open source [13], [14] and closed source [15], [16] software tools. After analyzing this literature, software components appear to be more similar to activity and time-based processes than hardware products. Therefore, partitioning (sequencing) analysis was chosen for the purposes of this paper. Sosa et. al (2007) analyzed the architecture of software products, by using clusters defined by the system architects and partitioned the

DSM using sequencing in order to uncover the dependencies that define the truly coupled components. The authors highlighted that using a sequencing algorithm is appropriate in software because software products exhibit a significantly large proportion of uni-directional dependencies between components [14], [15]. DSM partitioning (sequencing) is used in this paper in order to eliminate or reduce the feedback marks that exist above the diagonal. Partitioning a DSM makes it a lower triangular matrix by reordering rows and columns to ultimately make it a lower triangular matrix. This analysis will be presented in the following section.

3 Using DSM to Analyze Software Project Management Antipatterns

3.1 Data Collection

The knowledge required to build a DSM can be collected using traditional data collection methods, such as questionnaires [10]. Data on software project management antipatterns can be collected through detailed analysis of documented antipatterns. Due to the problems that exist with antipattern documentation, only a small number of software project management antipatterns authors publish antipatterns in books, research papers and on the Internet at the moment. In this paper, document analysis was used in order to populate the DSM with reliable information regarding software project management antipatterns. For the purposes of this paper, data on 16 antipatterns were collected. 13 antipatterns were collected from software project management antipattern books [1], [2], [3], private correspondence [4] and 3 antipatterns were collected from 3 software project management blogs[2] [3] [4] Table 1 contains a detailed description of the data used in this paper.

Antipatterns are documented using templates [2], [3] that consist of attributes that provide the structure of an antipattern. In this paper, the proposed framework considers antipatterns to be related through their causes, symptoms and consequences (see Table 2) [4]. Document analysis indicated 25 relationships between the 16 software project management antipatterns either through their causes, symptoms or consequences.

The collected data were used in order to build the DSM. Figure (1) illustrates an example of 3 antipatterns used for the creation of the DSM. This example [4] illustrates how antipatterns can be connected through their causes, symptoms or consequences. There exist 7 relationships among these attributes, which were taken into account in the design of the corresponding DSM model (See Fig. 2). The interpretation of the corresponding DSM model is described in the following subsection.

[2] http://blogs.msdn.com/nickmalik/archive/2006/01/03/508964.aspx

[3] http://blogs.msdn.com/nickmalik/archive/2006/01/21/515764.aspx

[4] http://leading_answers.typepad.com/leadinganswers/files/the_pipelining_antipattern.pdf.

Table 1. Sources of Software project management antipatterns

Source	Antipattern	Quantity
Books	Headless Chicken, Leader not manager, Road to nowhere, Metric Abuse [2] The Blob, Cut-and-Paste Programming, Reinvent the Wheel, Death By Planning, Project MisManagement, Stovepipe Enterprise, Stovepipe System [1] Planning 911, Architecture by Implication [3], [4]	13
Software Project Management Blogs	Project Managers who write specs, Guesses for Estimates , The Pipelining Anti-pattern	3

Table 2. Attributes relating software project management antipatterns

Causes	A list which identifies the causes of this antipattern.
Symptoms	A list which contains the symptoms of this antipattern.
Consequences	A list which contains the consequences that result from this antipattern.

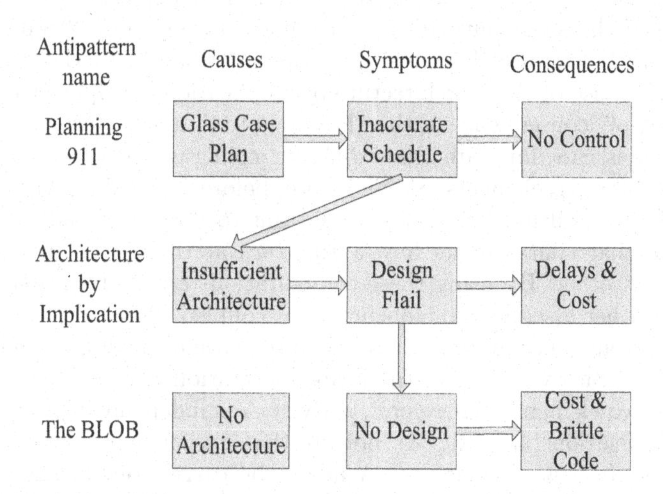

Fig. 1. Example antipattern attribute data [4]

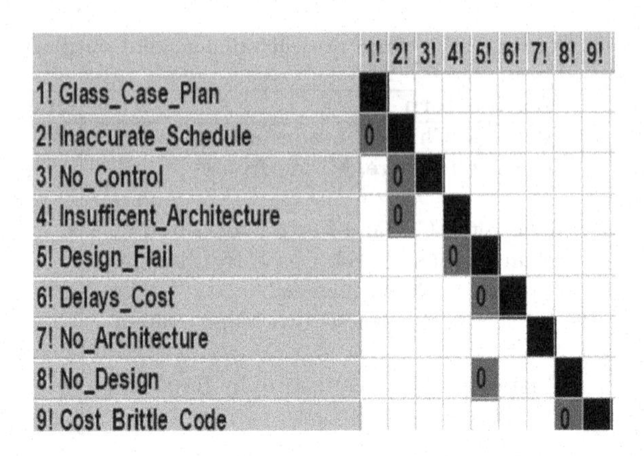

Fig. 2. Corresponding DSM of example data

3.2 DSM Analysis

DSM is a square matrix that shows relationships between elements (See Fig. 2). The quantification scheme in this paper employs the off-diagonal mark zero to indicate dependence between two antipattern attributes. In DSM, the row and column labels are identical but different information can be obtained by reading across a row and a column [8]. Reading across a row reveals other antipattern attributes that an attribute depends on. By reading columns, one can reveal what other attributes, an antipattern provides information to through its attributes. For example, in Figure (2), by reading across row 2, the symptom "Inaccurate Schedule" of "Planning 911" antipattern (See Fig.1) depends on the "Glass Case Plan" cause of the same antipattern. Reading across column 2 indicates that the symptom "Inaccurate Schedule" provides information to the "Insufficient Architecture" cause of the "Architecture by Implication" antipattern [3] and to the "No Control" consequence of the "Planning 911" antipattern.

DSM analysis can illustrate three different configurations that characterize relationships between elements [8]. Therefore, before proceeding with the interpretation of the resulting DSM, it is important to introduce and explain each configuration that characterizes how antipattern attributes can be potentially related to each other. By using these configurations antipattern attributes are described in either a sequential, concurrent or coupled fashion [8].

In a concurrent relationship (Fig. 3), two or more elements are independent and do not interact with each other. No information needs to be exchanged between the two elements. Therefore , activity A is independent of activity B.

Sequential relationships (Fig. 4) indicate that two or more elements depend on each other. In this case element A has to be carried out before element B can start because element B provides something to element A [18]. Therefore, element A affects the behaviour of element B in a uni-directional fashion.

The coupled configuration (Fig. 5) illustrates that two elements are interdependent. Element A and B both influence each other. This indicates that element

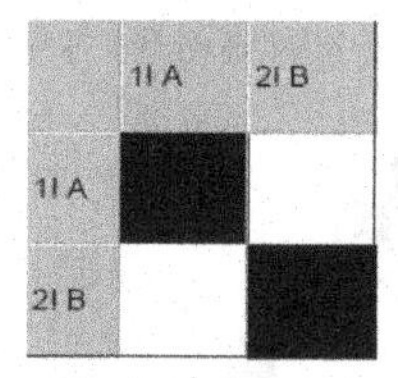

Fig. 3. Concurrent relationship DSM representation

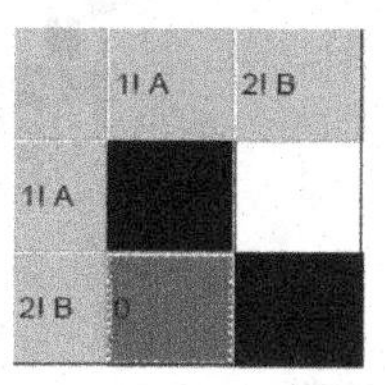

Fig. 4. Sequential relationship DSM representation

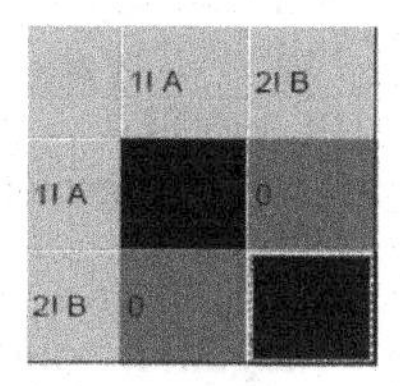

Fig. 5. Coupled relationship DSM representation

A cannot be carried before element B and likewise element B cannot be carried out before element A because they both provide something to each other. Changing one element will affect all other elements in such a block of interdependent elements. Choosing any element in such an iterative block or "circuit", one can follow the dependencies to any other element in that block and back again [9]. This kind of configuration between elements makes it difficult to decide which attributes must be addressed first. These "circuts" can be unwrapped by using assumptions in order to make "tears" [18]. The use of tearing in the antipattern attribute DSM is exemplified in the following section.

4 Data Analysis and Interpretation

The antipattern attribute DSM consists of information regarding relationships between 25 related attributes of 16 software project management antipatterns. The antipattern attributes in the unpartitioned DSM (Fig. 6) appear in a specific sequence. According to this sequence, the most important attributes appear in the top rows with the least important attribute at the last row of the DSM. Figure (6) illustrates the unpartitioned antipattern attribute DSM.

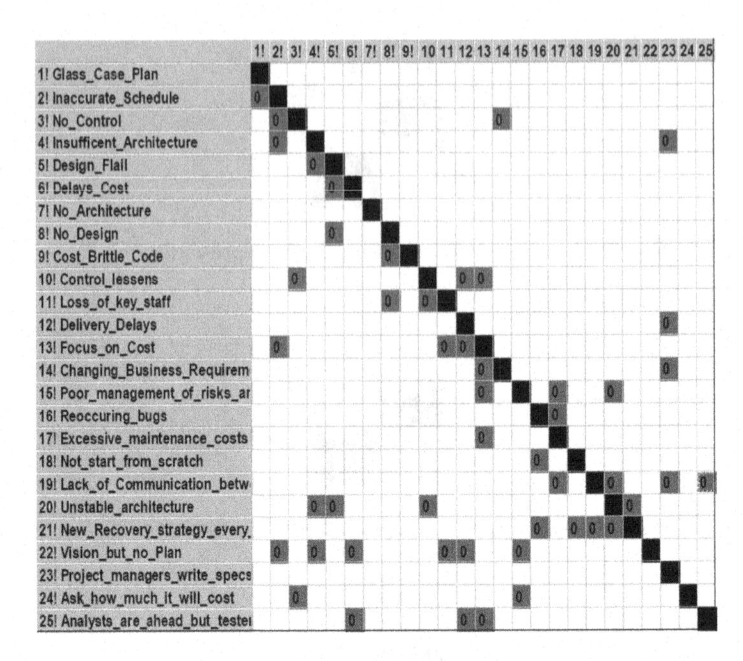

Fig. 6. Unpartitioned software project management antipattern attribute DSM

There exist 13 marks above the diagonal, which DSM considers as feedback marks [18]. This indicates the need to resolve interdependence in the antipattern attributes that lie above the diagonal. For example, reading across the fourth row of the unpartitioned matrix indicates that attribute "Insufficient Architecture" depends on information from attribute "Project Managers write specs" which is attribute number 23. This feedback mark implies that by the time "Insufficient Architecture" antipattern cause appears in a software project, it may have occurred because of an antipattern attribute that has not been resolved yet. This introduces the obvious need to address "Project Managers write specs" first as it is illustrated in the partitioned DSM (Fig 7). Therefore, DSM partitioning can be applied in order to eliminate feedback marks in antipattern attributes that appear above the diagonal by re-sequencing the attributes in the DSM. This approach can be used by software project managers in order to resolve antipatterns in an appropriate order.

PSM32 allows the automatic partitioning of sequential and concurrent elements in a DSM. It also allows users of this software to address interdependent (coupled) antipattern attributes by unwrapping dependency circuits. Figure (7) illustrates the partitioned antipattern attribute DSM. As already mentioned, DSM partitioning (sequencing) is used to eliminate or reduce the feedback marks that exist above the diagonal. In case all elements cannot be moved below the diagonal and become feedforward elements, then they are brought as close as possible to the diagonal. This ensures that each antipattern attribute will be addressed after all other required antipattern attributes have been addressed. By

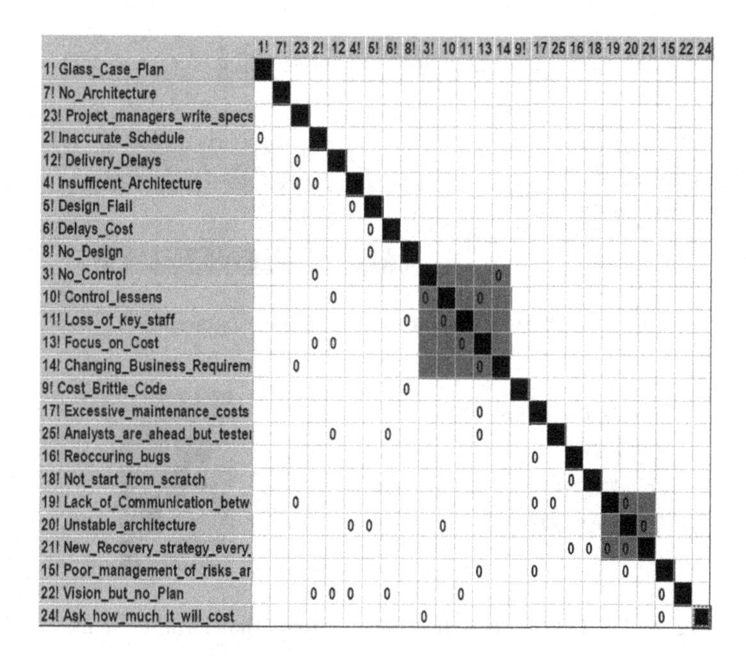

Fig. 7. Partitioned software project management antipattern attribute DSM

partitioning the antipattern attribute DSM (Fig. 7) using PSM32's partitioning function, only 4 out of the 13 feedback marks remain above the diagonal. All of these elements are part of interdependent coupled information circuits. The elements of the antipattern attribute DSM exhibited a significantly large proportion of concurrent dependencies. This indicates that a large number of antipattern attributes in the dataset used in this paper, are independent and can be addressed at any time in a software project.

The partitioned DSM revealed 4 sequential relationships that indicate uni-directional dependency. For example, attributes 16: "Reoccuring bugs" and 18: "Not start from scratch" depend on each other because "Reoccuring bugs" has to addressed before "Not start from scratch" is addressed. This occurs because according to the partitioned DSM (Fig. 7), "Not start from scratch" depends on "Reoccuring bugs" attribute in a uni-directional fashion.

DSM partitioning methodology can be used in order to unwrap the two information circuits that exist in the DSM. The first circuit appears in the highlighted rows between attributes 3 and 14 and the second circuit between rows 19 and 21. DSM tearing chooses the feedback marks (above diagonal) that if removed from the matrix, the matrix will become lower triangular when re-partitioned [18]. The marks removed from the DSM are called "tears". In order to discover the appropriate tears that will lead to a lower triangular matrix, assumptions are used in order to tear the use of an information item by temporarily removing its mark from consideration. Software tools can provide with assistance on deciding

Tear a column to move it to the bottom, a row to move it to the top.

Column Row PC Level: 1 2 3 4 5 6 7 8 9

Principal Circuit is 5 long	[x]	B	E	P	S	F	WS
14! Changing_Business_Requirements	[]	0	0	1	1	3	,600
3! No_Control	[]	0	0	1	1	3	,600
10! Control_lessens	[]	0	1	0	0	0	,000
11! Loss_of_key_staff	[]	0	0	0	0	0	,000
13! Focus_on_Cost	[]	1	0	0	1	3	,000

Fig. 8. PSM32 Tearing advice to unwrap circuits

14! Changing_Business_Requirem	0											0+
3! No_Control		0						0+				
10! Control_lessens			0						0+			0
11! Loss_of_key_staff							0			0+		
13! Focus_on_Cost		0	0								0+	

Fig. 9. PSM32 Marks indicating the items that form a circuit

where might be the best places to introduce the assumptions, i.e., make tears. PSM32 was used to make tears by using assumptions.

In order to exemplify this approach, the circuit between rows 3 and 14 shall be unwrapped using DSM methodology. By clicking on the first circuit consisting of 5 attributes, the tearing advice window of PSM32 (Fig. 8) can guide software project managers through the tearing process. This can provide details about the circuit that one wishes to tear as well as advice on where to tear. The rows appear in the order in which one goes around the principal circuit. The circuit is completed by going from the bottom item on the list back up to the top item.

PSM32 marks the elements that form the circuit with "+" signs (Fig. 9). The circuit is the following: 14 goes to 3 goes to 10 goes to 11 goes to 13 and goes back to 14. To unwrap the circuit one must find somewhere to introduce an assumption so as to break the circuit [9]. An assumption tears the use of an information item by temporarily removing its mark from consideration. This implies that instead of using an item that had been determined previously, an assumption is used. PSM32's Tearing Advice can be used to find the places to introduce make tears (assumptions). If a tear is made between 13 and 14 or between 14 and 3 a circuit remains, i.e. 3 goes to 10 and 10 goes back to 3. If a tear is made between 11 and 13, corresponding to the cell with the number 11 in the matrix, in this case with one attempt, all circuits are broken. This indicates attributes 10 and 11. PSM32 offers further tearing advice 8 by proposing that the best tearing result is obtained by choosing to tear a row where WS is .000. If more than one row has WS = .000, then the row where S is lowest should be torn. This also indicates attributes 10 and 11. Figure (10) illustrates the circuit with the column of attribute 11 marked as +5.

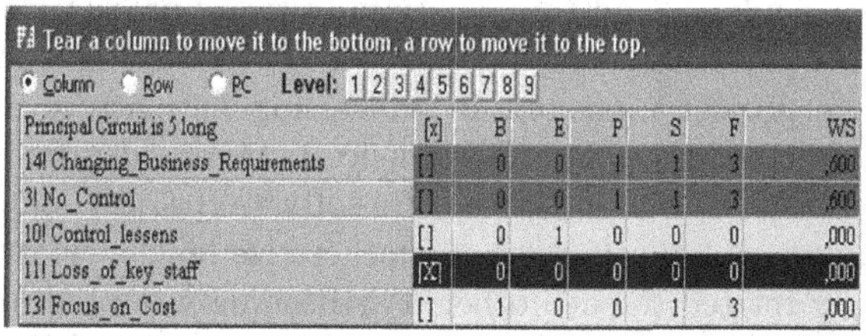

Tear a column to move it to the bottom, a row to move it to the top.

Column Row PC Level: 1 2 3 4 5 6 7 8 9

Principal Circuit is 5 long	[x]	B	E	P	S	F	WS
14! Changing_Business_Requirements	[]	0	0	1	1	3	,600
3! No_Control	[]	0	0	1	1	3	,600
10! Control_lessens	[]	0	1	0	0	0	,000
11! Loss_of_key_staff	[X]	0	0	0	0	0	,000
13! Focus_on_Cost	[]	1	0	0	1	3	,000

Fig. 10. Tear made in the column of attribute 11

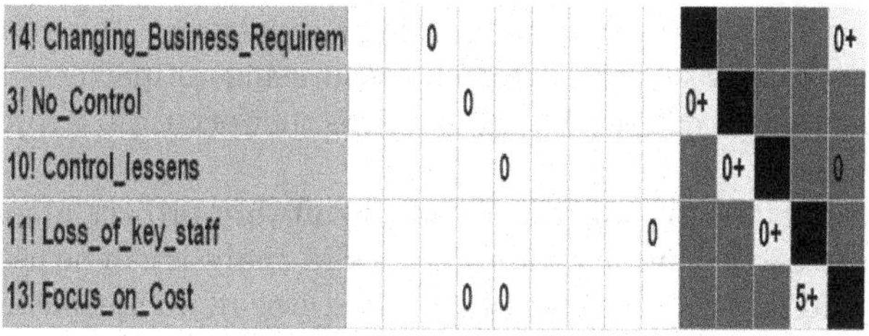

14! Changing_Business_Requirem			0											0+
3! No_Control				0						0+				
10! Control_lessens					0						0+			0
11! Loss_of_key_staff									0			0+		
13! Focus_on_Cost				0	0								5+	

Fig. 11. Tear made in attribute 11 using tearing advice

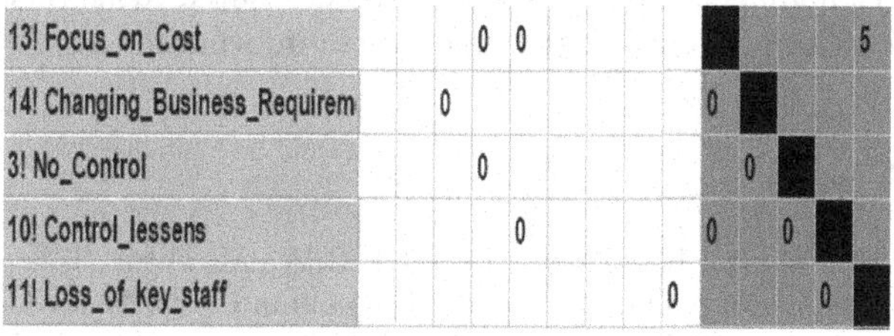

13! Focus_on_Cost				0	0									5
14! Changing_Business_Requirem			0							0				
3! No_Control				0							0			
10! Control_lessens					0					0		0		
11! Loss_of_key_staff									0				0	

Fig. 12. Partitioned DSM resolving the circuit

By placing a mark on attribute 11 (Fig. 11), every mark in the chosen column within the circuit is replaced with the chosen number.

By repartitioning the DSM the effect of this tear is illustrated in Figure (12). The circuit between interdependent antipattern attributes has been torn, leaving the only mark above the diagonal within the circuit and where the tear was made. Since there are no 0's above the diagonal within the circuit, the tear successfully broke the circuit. If any blocks remain, they could be torn in a similar manner.

5 Conclusion

In this paper, DSM has provided software project management antipatterns with new representation and analysis options. DSM methodology was particularly useful and well suited to the domain of antipatterns because it explicitly addressed interdependence and complexity in software project management antipatterns that occur during a software project. The resulting DSM can be used to assist software project managers in addressing antipattern attributes in a specific sequence that reduces interdependence between antipattern attributes.

The proposed framework can be used by software project managers to illustrate three different configurations that characterize how antipattern attributes can be potentially related to each other. Partitioning was used in this paper in order to eliminate or reduce the feedback marks that exist above the diagonal. The partitioned DSM was further analyzed using tearing in order to resolve interdependent information circuits of antipattern attributes. This approach can be used in situations where managers want to reduce interdependence and complexity in software project management using antipatterns. As a result DSM will ultimately benefit software project management using antipatterns by proposing a specific sequence that can resolve antipatterns that occurr in a software project in a timely manner.

Further investigation of applying DSM in software project management antipatterns can be carried out in future work. DSM analysis can be used to manage the effects of change [19]. For example, if the status of an antipattern symptom has to be modified (i.e. after resolving an antipattern by applying its refactored solution), it would be possible to quickly identify all other antipattern attributes, which were dependent on that attribute. This would reduce the risk with which software project managers carry out their project related tasks (i.e. monitoring) when their decision making is based on obsolete information.

References

1. Brown, W.J., Malveau, R.C., McCormick III, H.W., Mowbray, T.J.: AntiPatterns: Refactoring Software, Architectures, and Projects in Crisis. Wiley Computer publishing, Chichester (1998)
2. Laplante, P.A., Neil, C.J.: Antipatterns: Identification, Refactoring, and Management. Taylor & Francis, Abington (2006)
3. Brown, W.J., McCormick III, H.W., Thomas, S.W.: AntiPatterns in Project Management. Wiley Computer publishing, Chichester (2000)
4. McCormick III, H.W.: Antipatterns. In: Private correspondence, Presentation material - 3rd Annual European Conference on JavaTM and Object Orientation, Denmark (1999)
5. Settas, D., Bibi, S., Sfetsos, P., Stamelos, I., Gerogiannis, V.: Using Bayesian Belief Networks to Model Software Project Management Antipatterns. In: Proceedings of the 4th ACIS International Conference on Software Engineering Research, Management and Applications (SERA 2006), Seattle, USA, pp. 117–124 (2006)
6. Settas, D., Stamelos, I.: Using Ontologies to Represent Software Project Management Antipatterns. In: Proceedings of the 19th Sofware Engineering Knowledge Engineering Conference (SEKE 2007), Boston, USA, pp. 604–609 (2007)
7. Steward, D.V.: The Design Structure System: A Method for Managing the Design of Complex Systems. IEEE Transactions on Engineering Management 28, 71–74 (1981a)

8. Browning, T.R.: Applying the design structure matrix to system decomposition and integration problems: a review and new directions. IEEE Transactions on Engineering Management 48(3), 292–306 (2001)
9. Denker, S., Steward, D.V., Browning, T.R.: Planning Concurrency and Managing Iteration in Projects. Project Management Journal 32(3), 31–38 (2001)
10. Sharif, S.A., Kayis, B.: DSM as a knowledge capture tool in CODE environment. Journal of Intelligent Manufacturing 18(4), 497–504 (2007)
11. Sosa, M.E., Eppinger, S.D., Rowles, C.M.: 'A Network Approach to Define Modularity of Components in Product Design. Journal of Mechanical Design (2007)
12. Eppinger, S.D., Salminen, V.K.: Patterns of Product Development Interactions. In: International Conference on Engineering Design, Glasgow, Scotland (August 2001)
13. Sosa, M.E., Browning, T., Mihm, J.: Studying the dynamics of the architecture of software products. In: Proceedings of the ASME 2007 International Design Engineering Technical Conferences & Computers and Information in Engineering Conference (IDETC/CIE 2007), Las Vegas, USA, September 4-7 (2007)
14. MacCormack, A., Rusnak, J., Baldwin, C.Y.: Exploring the Structure of Complex Software Designs: An Empirical Study of Open Source and Proprietary Code. Management Science 52, 1015–1030 (2006)
15. Sangal, N., Jordan, E., Sinha, V., Jackson, D.: Using Dependency Models to Manage Complex Software Architecture. In: 20th ACM SIGPLAN Conference on Object-Oriented Programming, Systems, Languages And Applications (OOPSLA), San Diego, CA (2005)
16. Sullivan, K.J., Griswold, W.G., Cai, Y., Hallen, B.: The Structure and Value of Modularity in Software Design. ACM SIGSOFT Software Engineering Notes 26, 99–108 (2001)
17. Pimmler, T.U., Eppinger, S.D.: Integration Analysis of Product Decompositions. In: Proceedings of ASME 6th Int. Conf. on Design Theory and Methodology, Minneapolis (1994)
18. Yassine, A.: An Introduction to Modeling and Analyzing Complex Product Development Processes Using the Design Structure Matrix (DSM) Method, Quaderni di Management (Italian Management Review), No.9, English translation (2004)
19. Jarratt, T., Keller, R., Nair, S., Eckert, C., Clarkson, P.J.: Visualization Techniques for Product Change and Product Modelling in Complex Design. In: Blackwell, A.F., Marriott, K., Shimojima, A. (eds.) Diagrams 2004. LNCS (LNAI), vol. 2980, pp. 388–391. Springer, Heidelberg (2004)

An Embedded System Curriculum for Undergraduate Software Engineering Program

Gang Shen

School of Software Engineering
Huazhong University of Science and Technology
Wuhan, China 430074
gang_shen@yahoo.com

Summary. There is an increasing demand for well trained software talents in embedded application fields. Because embedded systems usually have complex requirements, higher standards are set for qualified embedded engineers' technical and soft skills. Considering the challenges imposed to the embedded system education, and given the resource constraints in universities, in this paper, we propose a practical embedded system curriculum for software engineering undergraduate programs. This curriculum consists of software engineering fundamentals, domain fundamentals and embedded system core courses. The two-phase embedded system core courses are aimed at improving students' abstracting and hands-on abilities in solving embedded system problems. The contents of these courses are selected according to the requirements from industry on the qualifications of embedded engineers. The implemented curriculum provides an proactive learning setting accepted by students. Close cooperation with industry is essential to the successful implementation of this curriculum.

1 Introduction

The term *embedded system* represents a large family of dedicated computing systems as a component of a larger system or device. Embedded applications can be found extensively in manufacturing, defense, transportation, telecommunications, consumer electronics and entertainment. J. Turley reported that as many as 99% of all processors made per year were used in various embedded systems [1]. Due to their importance, embedded systems keep attracting more attention in recent years. As a demonstration, we found over 16 million related web pages on the search engine Google, by using the keyword Embedded Systems. Compared with other software applications, embedded system development is unique in requirements, design, maintenance, and business models, requiring more effort on physical dimensions, product cost, real-time deadlines, reliability and development life cycle. This makes the education of embedded software developers different from that for general computer software developers.

Conventionally, many embedded developers were trained in the major application fields of embedded systems, like industrial control, signal processing, etc. Embedded system related courses based on particular architecture, such as microcontrollers, single board computers, systems on chips, and digital signal

R. Lee (Ed.): Soft. Eng. Research, Management & Applications, SCI 150, pp. 219–232, 2008.
springerlink.com

processors can be found in many areas outside of software engineering. These courses, including distributed digital control, robotics, critical systems, computer peripherals, etc., usually have their roots in, and furthermore are confined to, their specific application areas. The fast growth of embedded applications has frequently initiated new subjects, for example, network security, low-power computing, and real-time computing. Recently, there has been an effort to set up courses handling embedded systems as a separate and unified subject [2, 3, 4, 5]. Researchers with Carnegie-Mellon University studied many embedded system related courses in the U.S. undergraduate programs and concluded that there should be significant training for hands-on development. They remarked that the embedded system education, like other techniques and applications, should evolve over time [2]. Pak and others found the discrepancy between the needs of the IT industry and the embedded system software education offered by universities. Consequently, they designed embedded system courses to integrate the needs of the IT industry [3]. A series of experiments were devised to introduce software and hardware knowledge used in embedded system development for software engineering students [4].

Similar to [4], this paper is concerned with the embedded system curriculum for undergraduate software engineering programs. In 1997, a joint task force was founded by IEEE and ACM to promote software engineering as a profession. In one resulting document, SEEK'04, guidelines on embedded system education were suggested. But as a system and application specialty, embedded system related contents are distributed in embedded and real-time Systems, avionics and vehicular systems, industrial process control systems, and systems for small and mobile platforms courses [6]. These abstract guidelines leave much room for embedded system curriculum development.

In this paper, we describe efforts made rather than pursuing the completeness of the knowledge body of embedded systems. We focus on students' problem solving abilities. As a result, many practical components are added to enable students to actively learn embedded development. Realistic industry-like development settings are used to help students follow all steps in development lifecycles, ranging from problem identification, fact finding, information collection, analyzing and comparing solutions, to fast solution construction with the use or reuse of third party software and existing artifacts. The proposed H shaped curriculum contains software engineering fundamentals, including computer architecture, operating systems, user interface design and computer networks. It also has some basic electives in application fields, such as electronics, signals and systems, digital control, and information security. Linking the above two parts is the core embedded system knowledge, consisting of two courses. First, embedded system analysis and design that prepares students with the fundamental knowledge and analytical skills. Second, a comprehensive embedded system development project that helps students construct correct vision and useful hands-on experience. We stress the importance of cooperation with the industry that provides up-to-date development platforms, tools, cases and also the background for the project in the second course.

The rest of this paper is organized as follows: we first list the challenges and constraints that embedded system education has to face, and in Section 3 we discuss the requests from hiring companies for embedded system graduates, and subsequently use these requests as basis to design the problem-solving ability based curriculum in Section 4. Finally, we introduce and evaluate the implementation of this curriculum in Huazhong University of Science and Technology before presenting our conclusions.

2 Challenges and Constraints

Similar to many other developing fields, software engineering education is being tested by the continuous progress of technologies, and this is especially remarkable in embedded systems teaching, due to embedded systems' close connection to practical applications.

First, the advent of new hardware stills follows Moore's law, requiring constant course updating. For example, 4-bit microprocessors are still in use today in some devices, though the lowering cost of multi-core chips makes it feasible in embedded systems. All such things from specialized DSP to powerful generic PC, from EPROM with little capacity to large storage with cache, from industrial field buses to wireless sensor networks, can find their place in embedded applications. However, it is impossible to cover this huge landscape in a few courses taught within a limited time span. Of course, programmable logic controllers (PLCs) and DSPs play different roles in industrial control and telecommunications. For software engineering students, we must determine the goals of embedded system education to tailor the candidate materials.

Second, the complexity of embedded applications is growing. In earlier days, an application had only one task: control was switched between interrupt service routines and the main program. Multitasking, frequent interactions with other systems, and flexible operation flows make applications increasingly complicated. Decisions have to be made on whether to use the off the shelf commercial real-time operating systems to make development more efficient, or to create an entirely homemade system starting from the scratch to harness performance and reduce overhead. Another issue is that higher system complexity makes thorough testing unrealistic. To meet the quality goals, developers need to consider using verification and validation tools, or enforcing disciplined development processes.

Next, teamwork and communication training is a link missing in many existing embedded development curricula. Conventionally, the sole objective is to train individual's technical skills. But in our view, soft skills play an important role in the whole career of an embedded engineer. An embedded system developer may have to elicit requirements from customers and present solutions to peers and supervisors. Not any single meaningful software can be created by one person today. Therefore, an embedded system curriculum should cover training for both technical and soft skills.

In addition, there are a variety of development tools and technical variations for embedded system development. Eyeing on the importance of the embedded development market, both platform and software makers put forward sophisticated hardware and software tools for testing, designing and debugging embedded systems. To climb the learning curve of the large number of tools on the market, students need to spend considerably long time. Reasonable assumptions should be made to determine what should and should not be taught in lectures.

Different from other business development, embedded system requires low power consumption, high reliability, long life times, short development period, and since its cost is multiplied by quantity, a developer must consider the optimization of his design to save hardware costs [7]. Since embedded system per se means a high degree of quality, a curriculum must be compatible to the most critical requirements.

Additionally, the university education environment adds a number of constraints on embedded system curriculum. The first comes from the background and knowledge structure of instructors. Many instructors received complete theoretical training, with PhD degrees in Computer Science or Electrical Engineering, investigating in certain academic areas, but lack of first-hand working experience in industry. Since embedded system development requires strong application know-how, instructors should gain hands-on development experience before they are capable of imparting practical knowledge to students. Second, there is strict time constraint for embedded system education. The total credit hours for undergraduate students exceed 1800 hours in four years, and there is little flexibility for the time used in embedded system courses. Also, the resources available for embedded system education are limited. All reasonable experiments and projects need hardware platform support, and to make students get correct cognition of the embedded system development, the capstone project should be realistically complex and takes relatively long periods to finish. Ultimately, what attracts students in universities is not only classrooms and labs, but consistently motivating students is another task for embedded system instructors. Because it contains many details and requires continuous attention, embedded system development may make students get bored and reluctant to explore. Once students have learned fancy tools to create some working pieces, they may addict to playing with the specifics of tools and sacrifice their interest in abstract issues like modeling and validating.

In short, we need to consider the above challenges and constraints to design a practical embedded system curriculum.

3 Requirements for Embedded System Training

Requirements for graduates' skills have immediate influence on the design of the embedded system curriculum. As a matter of fact, a lot of information on embedded engineer's qualification requirements can be found on search engines like www.google.com, and www.baidu.com or North American and Chinese job search agents like www.monster.com, and www.51job.com. We used the keywords Embedded Engineer to search on the above web pages and randomly chose 100 distinct jobs as samples. These jobs included diverse application fields,

Table 1. Requirements for Qualifications of Embedded Engineers

Degree	Bachelor's Degree
Major	Computer Science Computer Engineering Electrical Engineering
Experience	Participation in all development phases Quick learning of new techniques Familiar with embedded software/hardware environments
Technical skills	C/C++ programming Network protocols Real-time software development Basics of hardware, firmware Object-oriented concepts One or more RTOS Creation and maintenance of documents
Soft skills	Oral and written communication Self-directed with little supervision, self-motivated, self-organized Focus on deadline and details Able to work independently and in team setting

ranging from telecommunications, equipment manufacturing, application platforms, automobile electronics, and industrial control to defense industry. In general, junior jobs required little or no experience in the specific application area, but senior jobs usually asked for more years of experience and in-depth domain knowledge. Excluding the requirements on years of experience and special domain knowledge, we classify the qualifications into the following categories: degree, major, experience, technical skills and soft skills. Those items appearing in more than 50% of the jobs are listed in Table 1. As well as those listed, knowledge of microprocessor architecture, fluency in one or more assembly language, knowledge of digital signal processing, debugging tools and disciplined software process (e.g. CMM or CMMI) also appear frequently.

We observe that the industry puts more emphasis on development skills and experience concerned with embedded system as a whole. This actually evidences that embedded system can be treated as a unified subject in software engineering education. It is also worth mentioning that good soft skills add values to embedded system engineers in landing a job. This is one of the principles that we will follow to design embedded system curriculum.

In China, the relatively low labor cost, high quality and sufficient supply of the talent pool has made software outsourcing grow dramatically in recent years. It is now one of the biggest sources for embedded system developer demand. Currently, outsourcing from Japan accounts for 60% of the total revenue (Ministry of Information Industry Annual Report 2006). According to a survey by one of our Japanese partners, main requirements for software engineering graduates

Table 2. Comparison of Software Development Cultures in China and Japan

Area	Category	Chinese	Japanese
Thinking	Personalities	Respect individuals	Group interest first
	Communication	Clarify personal points	Expect understanding from listeners
	Agreements	Reached by discussion	By consensus
Business	Doing business	Short term contract	Focus on long term collaboration
	Fulfilling contract	Stick to contract	Contract is paperwork, understanding is more important
Engineering	Information collection and exchange	Individual and reactive	Proactive as per need
	Programming style	Artistic	Standards followed
	Requirements changes	Reluctant and Hostile	Understandable
	Coding	Follow specifications	Do what should be done

are concerned with soft skills (see Table 2, source: Hitachi Systems GP). It is reported that Chinese software engineers are usually strong in techniques, but weak in foreign languages, slow to learn new knowledge, poor in making plans, lack skills in industrialized engineering environment, low in creativity and poor in communication. This reveals the fact that soft skill training is far from satisfactory in conventional university software education.

The fast growth of China's software industry has doubled its practitioners from 2000 to 2005. Among them, most are new graduates. The growth of the software industry needs more entrepreneurs and managers with creativity and leadership. Hence, leaving room for students with innovation and leadership development opportunities in embedded system courses, will add competitive advantage to students. From 2002, more than 30 state pilot software schools have been established to train more Chinese software talents, in order to meet the need from the rapidly expanding software industry. This makes exploring effective embedded system curriculums important and urgent.

4 Problem-Solving Based Curriculum Design

In the context of software engineering education in university, considering both the challenges from technical advancement and requirements on graduates' qualifications, we think, an embedded system curriculum should include both technical and soft skills critical to successful embedded system development, and this curriculum should encourage the construction of correct cognition from hands-on experience. To achieve these objectives, some principles are followed, i.e., active learning should play an important role, and enduring fundamental knowledge should dominate tool specifics.

Students learn better when they are more involved in lectures [8]. Due to the fact that embedded systems are evolving drastically year by year, practitioners have to have a proactive attitude and the ability to quickly learn new techniques and new tools to keep competitive in their careers. Then, it is helpful to foster active learning and quick learning habits and skills at embedded system classrooms and labs. Restricted by time limitation, a couple of embedded system core courses are unable to cover all the specifics of different hardware platforms and software utilities. Even if they could, there is no guarantee that the knowledge taught would not be obsolete after students start their jobs. We notice that the manuals of tools and software packages can be readily found from supplier's web pages or from developer's communities. When no clear and firm answer is available, students are encouraged to interact with the subject more proactively and generate their own knowledge, instead of hearing and then forgetting what they passively receive from the instructor. If students succeed in learning new techniques on their own, their confidence and motivation in learning more embedded system related knowledge will improve. It is fair to say that a well designed embedded system curriculum can provide a reasonable ambience for students to actively learn by hand-on training.

We argue that instructors should spend more time on the fundamentals instead of particularities of a tool, which should be left to students to learn from practice. The rationale is that the abstract concepts, reasoning, principles, methodologies, and formalism takes longer to understand, requiring systematic presentation, but their importance and value will benefit the learners long term. We suggest instructors spend more time elaborating the theoretical aspects on embedded systems and facilitate the active learning of other materials. What an instructor may do is first walk the students through one exemplar learning cycle of a typical tool, including using open resources to gather study materials, and then motivate students to learn the rest. As we have found, students are willing to learn the practical techniques, as long as they are able to see visible outcomes within short time.

Considering the diversity of application domains, between software and domain courses, we prefer to have more software engineering content. However, a reasonable proportion of domain courses should be open to students to select, preparing them with the fundamental understanding of the application background and laying some ground for them to master in-depth domain knowledge after their professional field is determined.

In the curriculum shown in Figure 1, software engineering fundamentals consist of programming languages, computer architecture, software architecture, object oriented analysis and design, software testing, user interface design, operating systems, computer networks and software project management, etc. These courses are slightly modified with a certain amount of embedded related materials added. The domain fundamentals comprise of signals and systems, electronics, digital control, information security and mobile computing etc. The

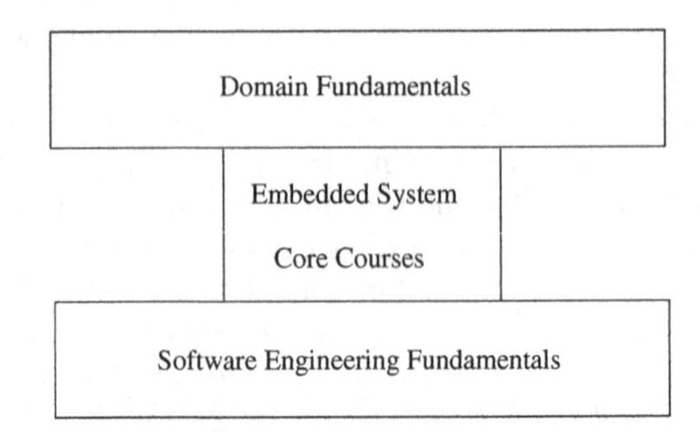

Fig. 1. An embedded system curriculum

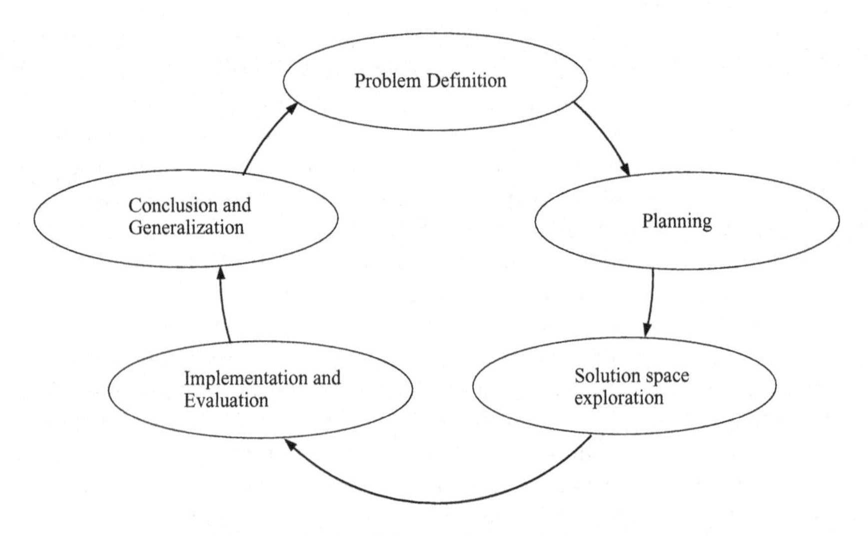

Fig. 2. A simplified problem-solving process

above two areas are connected by two-phase embedded system core courses: embedded system analysis and design, and a comprehensive embedded system development project.

Software is a practical discipline that is focused on satisfying customer's requirements. As revealed by constructivist learning theory, pushing students in the active problem solving role helps learning both problem-solving techniques and specialized knowledge. We plan to enhance students' problem-solving skills in limited embedded system courses. A simplified model is used throughout embedded system courses to depict the steps in a problem-solving process, i.e., identifying and exploring problems, making plans, comparing and selecting solutions, constructing and evaluating solutions, and concluding and generalizing (see Figure 2). By the problem-solving setting, we arrange embedded system core courses to training students with both technical and soft skills needed in embedded system development. In particular, we stress the importance of the

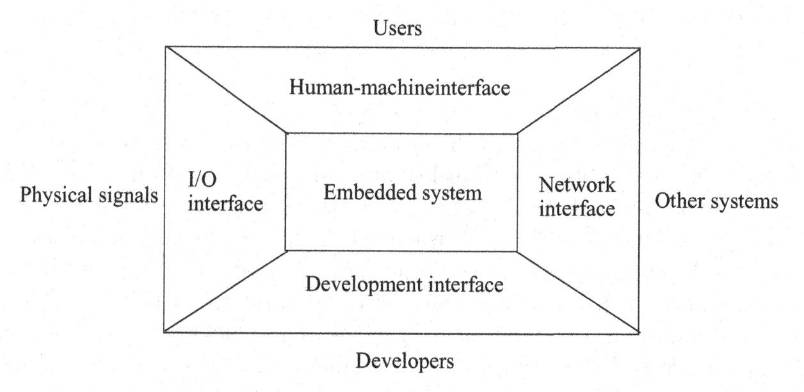

Fig. 3. Embedded system interfacing structure

planning phase and the generalization step in the problem-solving cycle, making students think and act in a systematic way.

The threads linking course disciplinary materials are three life cycles: embed system development life cycle, project management life cycle, and embedded program execution life cycle. Some subjects are recurring issues from prerequisite software engineering courses. For them, we stress on the uniqueness of embedded systems while linking them to more generic cases. From the perspective of problem-solving, innovation is equivalent to solving the problems of discovering and satisfying new requirements, adding new features, or improving existing applications with better performance or lower cost. In embedded system core courses, students are guided to follow the problem-solving process, using methodologies and knowledge taught in software and domain areas. To be specific, we use the interface structure in Figure 3 to illustrate that these interfaces are source of innovations on embedded systems. In many textbooks, interfaces from embedded systems to developers are either missing or introduced very briefly. However, in practice, interfacing an embedded system to developers is vital to its diagnosis, testing, debugging and upgrading. Therefore, we ask students to think of their activities as product development in a software organization, rather than simply finishing the course assignment.

In the first phase *embedded system core course Embedded System Analysis and Design*, lectures are organized on the hardware platform, software platform, real-time programming, I/O programming and development processes for embedded systems. The lectures contain the abstract modeling, analysis, evaluation, optimization and validation, i.e., the conceptual, formal and theoretical parts for embedded systems, as specified by SEEK'04. This course is accompanied by several inspiring mini-cases, experiments and demonstrations, to help students picture how various components of an embedded system work. The first phase encourages students to learn two aspects of embedded system development: formal methodologies, and engineering nature. In embedded development, planning, measurement, validating and making assumptions are essential subjects that will be repeated in the next phase. In the second phase, started in the following semester, the basics learned before will be used to solve practical problems: developing a working embedded system in a very demanding time frame.

In the course *Comprehensive Embedded System Development Project*, a capstone project requests students to first build up teams of 4 to 6 people. Each team has one student voted as team leader, and another one appointed as the on-site customer. Roles should rotate at least once in the middle of the course within a team. Each team will go through all development stages and manage the project independently. The project reflects one real-world embedded application need. There will be no absolute standard solution to the project, even no specific instructions available for operations. The instructor only specifies a business objective and grading criteria, providing basic hardware platforms and minimal software resources, leaving each team to set and adjust the project scope using their judgments. The instructor can be viewed as a tour guide, organizing students, making rules, arranging itinerary, leading students to different scene. Nonetheless, appreciating and exploring the scene are each team's responsibilities. As a supervisor and arbitrator, the instructor will set up milestones and checkpoints, and step in to in time feedback whenever the team deviates badly from the roadmap. The instructor takes more responsibilities at the last step of the problem-solving process, to summarize and generalize students' observations in the process.

As discussed before, soft skills, including solving problem independently, working efficiently in a team, understanding systematic development process, handling pressure, and communicating effectively are the keys to a successful career as an embedded engineer. In the proposed curriculum, soft skill improvement is not a secondary issue but a central theme. In the first phase of the embedded core courses, the objective of problem-solving in the second phase is announced early to make students aware of what they should learn in two courses taught within one year. In the second phase, we attempt to create a realistic practical development environment enabling students to actively learn the concrete facets of the abstract modeling, analysis, design and validating theories previously learned. The instructor shall make available major references on hardware/software platforms used in development. This capstone project is meant to simulate a real-world environment that students will face in the future, to train students to work in teams, under time pressure, with minimal supervision, on reasonably complex embedded system development.

In summary, our curriculum is aimed at training software engineering students on both technical and soft skills needed by system level embedded development. Restricted by time, the two-phase embedded system core courses are taught in two consecutive semesters respectively. The second module, a capstone project creates a realistic embedded systems development environment, helping students review what has been learned in the previous phase, and enable them to quickly learn the detailed use of new hardware and software tools. It is hoped that this project enables students obtain soft skills like team work, handling time pressure, as well as communication and presentation skills.

5 Implementation and Evaluation

In the software school of Huazhong University of Science and Technology, we have strong connection and active cooperation with IT industries, playing significant roles in embedded system teaching. In the first place, thanks to the donation from Intel, students have the opportunity to do experiments using XScale development kits. We also developed multi-core contents in advanced embedded system courses with the help from Intel. Sun Microsystem worked with us on operating system course development. The complete set of OpenSolaris course materials gives students good reference to learn details with respect to operating system implementations. Symbian makes possible students play with interesting embedded development on mobile platform. More importantly, industry provides us with solid application background of embedded development.

In the course Embedded System Analysis and Design, course objectives are introducing systematic approaches to developing embedded system software and best practices in managing embedded development project. In 32 hours, the following subjects are covered:

- Properties of embedded systems and embedded software development;
- Embedded hardware platform: ARM processor and its assembly language memory hierarchy, buses, I/O devices;
- Embedded software platform: RTOS basics, timing issues, scheduling algorithms, device drivers;
- Modeling: UML, state machines, task models, inter task communications, priority inversion, complexity and optimality;
- Verification, validation and simulation;
- Programming and debugging: system initialization, POSIX APIs, GDB, logging, task housekeeping;
- Embedded project: estimating and planning, prototyping, quality assurance.

Assignments and experiments are implemented on XScale development kits or on Linux PC.

One capstone project was originally a real project. With careful modifications, it becomes the background of the second phase embedded system course. In this project, an industrial control system consisting of a controller and a management station connected via Ethernet is to be developed. Generic personal computers (PC) and data acquisition cards (DAQ) are used in the project. This is partly because the number of hardware platform needed should match the number of student teams and also because there are plenty of available open source tools for PCs and DAQ cards. Students will develop a software programmable logical controller (PLC) and a software PID controller with the system architecture as shown in Figure 4. This system consists of fast periodic tasks for PID and PLC scans, slow periodic tasks for state and statistics update, and asynchronous tasks for user operations and network communications. The students have flexibility in defining their own characteristics of the user interface and PLC editor. The course instructor provides C programming conventions, templates for various documents, related web addresses for Linux, GTK, Comedi and RTAI references.

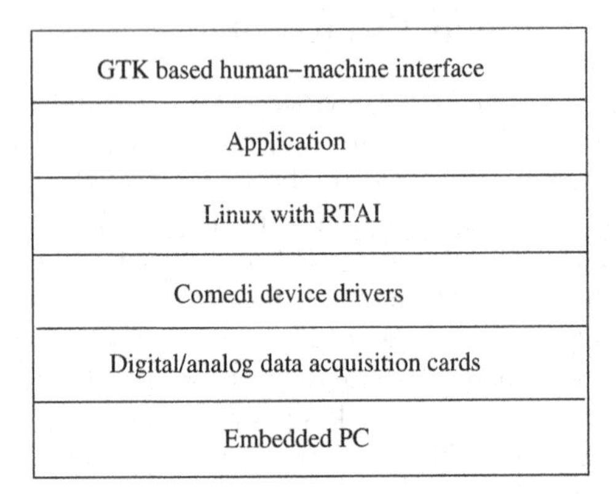

Fig. 4. A typical embedded system architecture

In this project, students need collect and analyze requirements, make project plan, design software architecture, design user interface, going through coding, debugging and testing, and integrating with the third party software (managing station). After all these are done, students submit deliverables, presenting and promoting the finished products team by team. In addition to those subjects in embedded courses, students are guided to connect their work with other software engineering fields, for example, using model/view/control pattern to design the software PLC. As optional subjects, students are encouraged to learn more in-depth techniques and do serious research given this application background, possibly as part of their diploma thesis work. An example is to explore the use of reduced ordered decision diagrams (ROBDD) for the software PLC in design. As a compact representation of Boolean functions, ROBDDs have been introduced as a model checking tool in the previous course, however, students have to make further investigation to find a good model for PLC that is able to facilitate both edition and execution.

After the project, almost all students reported that in addition to the improvement of technical skills, they share some points of view on soft skill gains:

- They understand the importance of team synergy. Because the project contains many aspects of embedded system development, it is a big challenge in terms of effort. An individual is unable to finish all tasks. A team can succeed only after its members communicate effectively;
- They understand the time pressures in embedded system development. The average real embedded development period is 6 to 8 months long. This project needs careful scheduling and all team members must stick to this schedule. Testing and debugging takes a lot of time. Many defects are detected after integration.
- This cultivates a feeling of pride and confidence. They are able to complete a working system and no longer fear hands-on development.

We observed that several problems should be handled in the further development of this curriculum. The participation is not consistent among different team members. There is obvious disparity of individual contribution to the system developed. Even with the role rotation, under time pressure, some more competent team members would volunteer to take more and critical responsibilities. Since there is no real customer in the project, some students are used to thinking the instructor as their customer. When the instructor cannot provide timely information, such as specific requirement or reviewing code, they might tend to lose the quality control and lower the standards.

6 Conclusions

In this paper, we introduced an embedded system curriculum in a software engineering undergraduate program. As a promising application field, embedded systems need a lot of well trained software graduates. Because embedded systems have more demanding requirements, compared with other software applications, higher standards are needed for embedded engineer's technical and soft skill sets. We analyzed the development trends of embedded systems and hence the challenges they impose to embedded system education. Given the restrictions of resources available in university, we designed an embedded system curriculum consisting of software engineering fundamentals, domain fundamentals and embedded system core courses. The two-phase embedded system core courses are aimed at improving students abstracting and hands-on abilities in solving embedded system problems. Following the requirements from the industry on embedded engineers' qualifications, we arrange the contents of these courses. The implemented curriculum provides students with practical active learning setting accepted by them. Close cooperation with industry makes up the deficiency of university's resources.

References

1. Turley, J.: Embedded Processors by the Numbers. Embedded Systems Programming 12(5) (May 1999)
2. Koopman, P., Choset, H., Gandhi, R., et al.: Undergraduate embedded system education at Carnegie Mellon. ACM Transactions on Embedded Computing Systems 4(3), 500–528 (2005)
3. Pak, S., Rho, E., Chang, J., et al.: Demand-driven curriculum for embedded system software in Korea. ACM SIGBED Review 2(4), 15–19 (2005)
4. Wang, H., Wang, T.: Curriculum of Embedded System for Software Colleges. In: Proceedings of the 2nd IEEE/ASME International Conference on Mechatronic and Embedded Systems and Applications, Beijing, August 2006, pp. 1–5 (2006)
5. Gross, H., van Gemund, A.: The Delft MS curriculum on embedded systems. ACM SIGBED Review 4(1), 1–10 (2007)

6. IEEE Computer Society/ACM Joint Task Force. Software Engineering 2004 - Curriculum Guidelines for Undergraduate Degree Programs in Software Engineering (2004)
7. Wolf, W.: Computers as Components: Principles of Embedded Computer System Design. Morgan Kaufman, San Francisco (2000)
8. Hall, S.R., Waitz, I., Brodeur, D., et al.: Adoption of active learning in a lecture-based engineering class. Frontiers in Education 1 (2002)

Model Checking for UML Use Cases

Yoshiyuki Shinkawa

Faculty of Science and Technology, Ryukoku University
1-5 Seta Oe-cho Yokotani, Otsu 520-2194, Japan
shinkawa@rins.ryukoku.ac.jp

Summary. A use case driven approach is one of the most practical approaches in object orientation. UML use case diagrams and their descriptions written in a natural language are used in this modeling. Even though this approach provides us with convenient ways to develop large scale software and systems, it seems difficult to assure the correctness of the models, because of insufficient formalization in UML. This paper proposes a formal model verification process for UML use case models. In order to exclude ambiguity from the models, they are firstly formalized using the first order predicate logic. These logic based models are then transformed in the form of Promela code, so that they can be verified using the Spin model checker. A Promela code must be composed based on state transitions, whereas the logic based use case models do not explicitly include the states and their transitions. Therefore we introduce a state identification process in the logic based use case models. A supermarket checkout system is used to show how the proposed process works.

1 Introduction

A use case driven approach provides us with a practical way to analyze complicated requirements or system behavior in large scale software developments [7, 10]. In this approach, use case modeling is the core activity, which creates many kinds of use case diagrams and their descriptions based on UML (Unified Modeling Language) [9, 4]. Each use case model represents a fragment of the system usage, and these fragments compose the external behavior of the system to be developed.

This approach is effective in the requirement analysis for large scale software or systems, since many stakeholders and interest groups with different knowledge, roles, or missions, can provide their operational or external requirements only from their own viewpoints without caring about the complicated whole system behavior.

However, these independently created use case models might include conflicts within or between them. Since the use case models form the foundation for the later software development activities, these conflicts or inconsistencies could cause severe problems, e.g. malfunction of systems, wrong outputs, or unimplementable specifications.

Therefore, it seems important to identify and exclude these conflicts. Many efforts have been made to formalize UML models in order to assure their correctness [8, 5, 12]. However there are still several difficulties in identifying the conflicts, since

1. few metrics are defined in UML use case modeling to evaluate the conflicts [3]

R. Lee (Ed.): Soft. Eng. Research, Management & Applications, SCI 150, pp. 233–246, 2008.
springerlink.com

2. UML use case models are very little formalized, which are described in the form of UML use case diagrams in conjunction with their descriptions written in a natural language [11]

In order to relieve the problems, we have to represent use case models more formally and rigorously, with explicit rules for model correctness and consistency.

This paper proposes a formal approach to evaluating the correctness of UML use case models, using logic based formalization and a model checking technique. The paper is organized as follows. Section 2 discusses how UML use case models are transformed into formal descriptions in the form of logic formulae. Section 3 introduces a model checking technique to evaluate the correctness and consistency of the models that are described as logic formulae. The Spin model checker [6, 2] is used as a model checking tool in this paper. Therefore, all the logic formulae are transformed in the form of Promela codes. In section 4, a systematic way to evaluate the above Promela codes using the Spin. LTL (Linear Temporal Logic) formulae that are used to evaluate the codes are derived from the original use case models.

2 Logic Based Formalization of Use Case Models

A use case model expressed by UML usually consists of two parts, namely a set of use case diagrams and a set of their corresponding descriptions. A use case diagram represents how people interact with a system in order to accomplish meaningful tasks. A use case diagram is composed of the following three basic model elements [1].

1. Use cases: A use case is a series of operations or *actions* for accomplishing some purpose, interacting with the system.
2. Actors: An actor represents a person, an organization, or a system that is related with the use cases.
3. Associations: An association exists between an actor and a use case, which represents that the actor is involved with the use case.

In addition to the above, two optional elements are defined for notational convenience in use case diagrams, namely system boundary boxes and packages.

While use case diagrams describe high-level or abstract-level interactions with a system, use case descriptions represent more concrete operations or actions residing within the use cases, that is, they provide us with more detailed information about the use cases. A use case description is a textual expression on a use case, which is composed of the following major sections.

1. Pre-conditions: A list of conditions that must be met before a use case is invoked.
2. Post-conditions: A list of conditions that will be true after the use case finishes.
3. Basic course of action: A sequence of actions that an actor follows, when no exceptional conditions occur.
4. Alternate course of action: An infrequently used sequence of actions, e.g. a sequence that is performed under some exceptional environments. This course replaces consecutive actions in the basic course. At the end of this course, it joins the basic course, or exits from the use case.

Table 1. Use Case Description

Name: Receive a Basket
Preconditions: A cashier is ready for service
Postconditions: All the items are unprocessed
Basic Course of Action
1. A customer arrives at the cashier 2. The cashier receives a new shopping basket
Name: Process an Item
Preconditions: At least one item is left unprocessed in the shopping basket
Postconditions: The item is treated as processed
Basic Course of Action
1. The cashier picks up an item 2. The cashier scans the barcode on the item 3. The product code of the item is sent to the store system to get the price 4. The price is displayed on the POS screen
Name: Payment
Preconditions: All the items in the shopping basket are processed
Postconditions: The cashier is ready for new customers
Basic Course of Action
1. The cashier shows the total amount to the customer 2. The cashier receives payment ⇒ (The alternate course A) 3. The cashier inputs the received amount through the POS keyboard 4. The POS printer prints out the receipt 5. The cashier gives the receipt with the change
Alternate Course of Action A: Payment by a credit card
A2. The cashier receives a credit card A3. The cashier scans it using a credit card reader A4. The card data is sent to an authorization computer A5. The cashier passes the bill to the customer A6. The customer signs the bill ⇒ The use case ends

In addition to the above, some supplementary information like "*name*", "*identifier*", and "*description*" may be included. In order to make use case models rigorous, we first define the constituents of use case models more explicitly. Since use case diagrams are simple figures, we focus on the constituents of use case descriptions for rigorization.

In use case descriptions, we have to deal with pre- and post-conditions that are characterized by boolean values, and action flows for basic and alternate courses of action, each of which perform some functionality.

As for conditions, logical formulae of first-order predicate logic can represent them more rigorously than textual expressions that are often used in use case descriptions. In order to compose the logical formulae, we need functions, predicates, variables, and a set of objects that are referred to in the functions and predicates, along with logical operators such as "$\vee$", "$\wedge$", "$\neg$", "$\forall$", and "$\exists$".

Even though logical formulae can represent pre- and post-conditions rigorously in comparison with textual notation, each formula may be expressed in many different

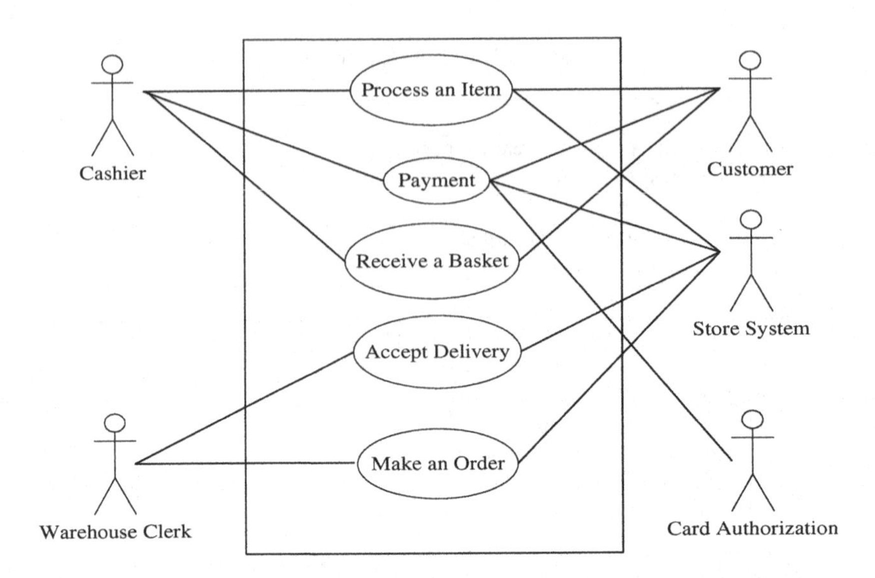

Fig. 1. Use Case Example

ways. This variety of notation makes it difficult to understand the conditions, and might lead us into confusion. In order to avoid such difficulty, we assume each condition is expressed as a "*prenex conjunctive normal form*" (PCNF), that is, it is written as a string of quantifiers ($\forall$ and $\exists$), followed by a conjunction of clauses.

Once all the pre- and post-conditions are expressed in the form of PCNF, the next step is to express basic and alternate courses of action rigorously. An action is the smallest unit of interaction between an actor and a system, which might affect some objects in the system. Each action can be regarded as a method at an implementation level, and therefore it is characterized by a method name and a signature.

These names and signatures will be extracted by parsing the textual expressions that occur in the courses of action, focusing on the subjects, verbs, and the objects of the verbs. As the result, we obtain a set of method or action names, and a set of object types that characterize the objects. An action is denoted $a_s(s = \sigma_1 \ldots \sigma_n, \sigma)$ using the above action names a and a series of object types s, where s is the signature of a. This notation of an action is identical to that of an S-sorted function in many sorted algebra. In many cases, two or more different actions with the same name and with the different signatures might be extracted from the textual descriptions, which are to be treated as the different actions. These actions represent polymorphic functions, and must be distinguished. We append an appropriate suffix to each action like $a_1, a_2, \cdots$ in such a case.

In order to show how the proposed formalization process works, we use a use case model for a supermarket checkout system, which is represented by the use case diagram in Fig. 1 and the use case descriptions in Table 1. Among the five use cases in Fig. 1, three selected use cases "Receive a Basket", "Process an Item" and "Payment" are represented in the form of use case descriptions.

Table 2. Formal Use Case Description

Name: Receive a Basket
Preconditions: $\exists e \in E[\text{isWaiting}(e)]$
Postconditions: $\forall s \in B_i[\neg\text{isProcessed}(s)]$
Basic Course of Action
1. $\text{arrive}(e)\quad e \in E$ 2. $d = \text{receiveBasket}(B_i)\quad d \in D,\ B_i \in B$
Name: Process an Item
Preconditions: $\exists s \in B_i[\neg\text{isProcessed}(s)]$
Postconditions: $\text{isProcessed}(s)$
Basic Course of Action
1. $s = \text{pickup}(B_i)\quad s \in B_i,\ B_i \in B$ 2. $p = \text{scan}(s)\quad p \in P$ 3. $price1 = \text{getPrice}(p)\quad price1 \in \text{int}$ 4. $price2 = \text{displayPrice}(price1)\quad price2 \in \text{String}$
Name: Payment
Preconditions: $\forall s \in B_i[\text{isProcessed}(s)]$
Postconditions: $\text{isReady}(d)\quad d \in D$
1. $totalAmmount = \text{showTotalAmmount}(B_i)\quad B_i \in B,\ totalAmmount \in \text{int}.$ 2. $payment = \text{receivePayment}(B_i)\quad payment \in \text{int}\quad \Rightarrow$ (The alternate course A) 3. $\text{inputPayment}(payment)$ 4. $receipt = \text{getReceipt}(B_i)\quad receipt \in \text{String}$ 5. $\text{giveReceiptAndChange}(receipt, change)\quad change \in \text{int}$
Alternate Course of Action A: Payment by a credit card
A2. $c = \text{receiveCreditCard}(B_i)\quad c \in C$ A3. $cardNumber = \text{scanCard}(c)\quad cardNumber \in \text{int}$ A4. $result = \text{cardAuthorize}(cardNumber)\quad result \in \text{boolean}$ A5. $bill = \text{printBill}(B_i, c)\quad bill \in \text{Strong}$ A6. $\text{signBill}(bill)\quad \Rightarrow$ The use case ends

From the given use case description, we can identify the following combination of the sets.

- A set of shopping items $S = \{s_1, s_2, \ldots\}$, each member of which represents a shopping item currently sold at the supermarket.
- A set of cashiers $D = \{d_1, d_2, \ldots\}$
- A set of customers $E = \{e_1, e_2, \ldots\}$
- A set of shopping basket $B = \{B_1, B_2, \ldots\}$. Since each B_i contains the shopping items chosen by a customer, $B_i \subset S$ holds.
- A set of product $P = \{p_1, p_2, \ldots\}$, each member of which is associated with a product code. A shopping item $s_i \in S$ is associated with one of the $\{p_j\}$.
- A set of credit cards $C = \{c_1, c_2, \ldots\}$.

Over these sets, the following functions or methods are used in the formalized use case descriptions

- arrive: $E \longrightarrow$, which starts a service (The function "arive" returns void)
- receiveBasket: $B \longrightarrow D$, which receives a shopping basket from a customer

- pickup: $B \longrightarrow S$, which takes an item from a shopping basket
- scan: $S \longrightarrow P$, which gets the product code of a shopping item
- getPrice: $P \longrightarrow$ int
- displayPrice: $P \longrightarrow$ String
- showTotalAmmount: $B \longrightarrow$ int
- receivePayment: $B \longrightarrow$ int (We assume the payment is made for each shopping basket)
- inputPayment: int $\longrightarrow$
- printReceipt: $B \longrightarrow String$
- giveReceiptAndChange: String $\times$ int $\longrightarrow$
- receiveCreditCard: $B \longrightarrow C$
- scanCard: $C \longrightarrow$ int
- authorizeCard: $C \longrightarrow$ boolean
- printBill: $B \times C \longrightarrow$ String
- signBill: $B \longrightarrow$

The predicates that would be used to express the above descriptions are

- $\text{isProcessed}(s) : s \in B_i$
- $\text{isReady}(d) : d \in D$
- $\text{isWaiting}(e) : e \in E$

Using the above constituents, the use case descriptions in Table 1 is expressed as shown in Table 2. Other descriptions can be expressed in the same way.

3 Transforming Logic Formulae to Promela Code

Even though the above logical formulae and several series of functions or methods can express the use case models rigorously, they can not prove their correctness and consistency themselves, since they are simply transformed from the original use case models that might include conflicts. We need external rules, metrics, and processes to evaluate the correctness and consistency of these models.

The Spin model checker is one of the most practical model checking tools to evaluate the correctness of models, based on state transition systems and LTL (Linear Temporal Logic). In order to use the Spin to evaluate the logic based use case models, the models must be expressed using Promela language that the Spin can deals with.

Before transforming the logic based use case models into Promela codes, we first define the basic structure of the logic based use case models to be evaluated.

3.1 Basic Structure of Logic Based UML Use Case Models

As discussed in section 2, each use case model M is expressed as a quadruple

$$M = \langle P, B, \{A_i\}, Q \rangle$$

where P is a pre-condition, B is a basic course of action, $\{A_i\}$ is a set of alternate courses of action, and Q is a post-condition. P and Q are represented as logical formulae in the form of PCNF. B is a series of functions $f_1, \cdots, f_m$, and A_i is a quadruple $\langle b_s, P_L^{(i)}, \{g_1^{(i)}, \cdots, g_{qi}^{(i)}\}, Q_L^{(i)} \rangle$, where the integer $b_s > 0$ represents the step number in the basic course, at which the alternate course branches, $g_j^{(i)}$ is a function for the corresponding action, $P_L^{(i)}$ is a pre-condition, and $Q_L^{(i)}$ is a post-condition for the alternate course A_i. As these conditions are defined locally within the alternate course, we refer to the conditions $P_L^{(i)}$ and $Q_L^{(i)}$ as *local conditions* henceforth.

These use case models are often created independently, however they could be interrelated, e.g. some use cases are the prerequisites or the postrequisites for other use cases. These interrelationships can be defined based on the pre- and post-conditions of the use case models. Each use case is activated when its pre-condition is satisfied, and its post-condition is satisfied after the use case is done. Therefore if the post-condition Q of a use case M is identical to the pre-condition P' of another use case model M', or more generally, if $Q \vdash P'$ holds, that is, P' is deduced from Q, the use cases M and M' are successively performed.

As discussed above, a use case model can have multiple post-conditions, one is for the basic course of action and the others are for alternate courses, while it can have only one post-condition [1]. This implies each use case model M can be expressed as a unit with one input port for the pre-condition, and multiple output ports for the basic and alternate courses of action. These use case models can be connected based on the following rules.

1. If the post-conditions $Q_1, \cdots, Q_n$ of the use case models $M_1, \cdots, M_n$ and the pre-condition P of the model M satisfy $Q_1, \cdots, Q_n \vdash P$, the output ports of $M_i (i = 1, \cdots, n)$ are connected to the input port of M. In this case, the use case M is performed after all the use cases $M_1, \cdots, M_n$ end.
2. If $Q_1 \vdash P, \cdots, Q_n \vdash P$ hold in the above case 1, the output ports of $M_1, \cdots, M_n$ are connected to the input port of M. This connection represents the "*or join*h. In this case, the use case M is performed after one of the use cases $M_1, \cdots, M_n$ ends.
3. If the post-condition Q of a use case model M and the pre-conditions $P_1, \cdots, P_m$ of the models $M_1, \cdots, M_m$ satisfy $Q \vdash P_1, \cdots, Q \vdash P_m$, the output port of M is connected to the input ports of $M_i (i = 1, \cdots, m)$. In this case, the use cases $M_1, \cdots, M_m$ are performed after the use case M ends.
4. The above post-conditions can be substituted by local post-conditions arbitrarily.

Using the above rules, we obtain a set of directed networks of use cases.

3.2 Defining State Transitions in Use Case Models

A logic based use case model is composed of logic formulae for pre- and post-conditions, and several series of functions for basic and alternate courses of action. On the other hand, a Promela code represents a state transition system, and is composed of a variable definition part and process definition parts. A state transition is implemented by assigning a value that represents a state to a variable.

[1] Even though a use case have multiple *local* pre conditions, they only are evaluated after the use case is activated, and never concern the activation of the use case.

Therefore, in order to transform a logic based use case model into a Promela code, we first have to define the states and their transitions in a use case model. Unlike a Promela code, variables in a use case model are not explicitly associated with states, and assigning values to the variables is implicitly shown in the form of function calls or method invocations.

In a use case model, the states of the model can be regarded as a set of states of the objects that occur in the model. The states of each object are changed when a method is invoked, which updates the object. The state of an object can be changed by its own method, by a method that receive it as an argument, or by a method that can access it implicitly.

Since the state of a use case model at a moment can be defined as a set of the states of these objects, the state transition of the model can be traced by examining the state transition of each object through the associated methods. This trace is performed in the following way.

1. Examine each object in the model to determine whether it affects the state of the model. If it does, we refer to it as a *state-sensitive* object.
2. For each state-sensitive object that is identified in the step 1, examine the methods or functions in the model to identify which of them can update the state of the object. These methods are referred to as *state-sensitive* to the object.
3. The state of each state-sensitive object is determined by a state mapping function $\mathscr{S}: \Omega \to \Sigma$, which maps from a set of state-sensitive objects Ω to a set of all the possible states Σ. This $\mathscr{S}$ is updated by the execution of a state-sensitive methods. The updated function by the consecutive execution of the methods $f_1, \cdots, f_n$ is denoted by $\mathscr{S}_{f_1 \cdots f_n}$, and the state of an object $x \in \Omega$ under this environment is denoted as $\mathscr{S}_{f_1 \cdots f_n}(x)$.

In a use case model $M = \langle P, B, \{A_i\}, Q \rangle$, there could be multiple different sequences of method invocations, depending on which course of action is selected. Assuming the above use case model M includes the state-sensitive objects $x_1, \cdots, x_n$, the basic course of action $B = f_1, \cdots, f_m$, and the alternate courses of action $A_i = \langle b_i, P_L^{(i)}, \{g_1^{(i)}, \cdots, g_{q_i}^{(i)}\}, Q_l^{(i)} \rangle (i = 1, \cdots, p)$, the possible state mapping functions are $\mathscr{S}_{f_1 \cdots f_m}$ and $\mathscr{S}_{f_1 \cdots f_{b_i - 1} g_1^{(i)} \cdots g_{q_i}^{(i)}} (i = 1, \cdots, p)$. If a method f_r is not state-sensitive to x_j, $\mathscr{S}_{f_1 \cdots f_{r-1}}(x_j) = \mathscr{S}_{f_1 \cdots f_r}(x_j)$ holds.

3.3 Transforming Use Case Models into Promela Codes

In order to evaluate a use case model by the Spin model checker, the above state transitions must be expressed in the form of Promela codes. Since Promela represents the states of objects as the variables and their values, we have to define the variables that correspond to the state-sensitive objects, and the values that determine their states.

While the variables can be simply mapped from the state-sensitive objects, their states are determined by which methods are invoked within the model. The state mapping function $\mathscr{S}$ that is introduced in Section 3.2 can be expressed in the form of $\mathscr{S}_{\varphi_1 \cdots \varphi_n}$ where φ_i represents a method that occur in the basic or alternate courses of action in the model. For each state mapping function, if $\varphi_l (l \leqq n)$ is state sensitive to the object x_j, $\mathscr{S}_{\varphi_1 \cdots \varphi_l}(x_j)$ is a new state that is caused by a sequence of the method invocations $\varphi_1 \rightarrow \cdots \rightarrow \varphi_l$. By assigning appropriate state names to these new states, we can define a set of states to be taken into account in the model.

A Promela code that reflects the above situation is constructed in the following way.

1. Define the variables $x_1, \cdots, x_n$ for each state-sensitive object.
2. Define **mtype** $= \{s_j^{(i)}\}(i = 1, \cdots, n)$, where $\{s_j^{(i)}\}$ is a set of possible states of the object x_i that are obtained by examining each state mapping function $\mathscr{S}_{\varphi_1 \cdots \varphi_m}$.
3. For each method φ_j, put the series of $x_i = s_j^{(i)}$ within an atomic block, if φ_j is state-sensitive to x_i, where $s_j^{(i)} = \mathscr{S}_{\varphi_1 \cdots \varphi_j}(x_i)$.
4. Compose an active process as a series of the above atomic blocks if there is no alternate course of action.
5. If there is an alternate course of action at the b_ith method of the basic course, put
   if
   ::(C) $\rightarrow$ atomic{the series of the methods for the alternate course}
   ::else $\rightarrow$ atomic{the series of the methods for the basic course}
   fi
   where (C) is the condition for the alternate course.

The above process can easily be extended to the interconnected use case models that is discussed in Section 3.1.

4 Evaluating Use Case Models Using LTL

In the Spin model checker, each Promela code is examined against the constraints that are expressed in the form of LTL (Linear Temporal Logic) formulae based on the states of variables. On the other hand, in our logic based use case models, these constraints are provided as pre- and post-conditions in the form of PCNF first order predicate logic formulae. Since LTL and first order predicate logic formulae are composed differently, in syntax and semantics, we have to transform the pre- and post-conditions into LTL formulae that the Spin can deal with. Before discussing this transformation, we first define the relationships between *predicates* and *states*.

The predicates that occur in pre- or post-conditions refer to the objects in the use case models. These objects $x_1, \cdots, x_n$ occur as the arguments of a predicate P in the form of $P(x_1, \cdots, x_n)$. This $P(x_1, \cdots, x_n)$ is evaluated *true* or *false* based on the values of $x_1, \cdots, x_n$. These values are regarded as the states of the objects $x_1, \cdots, x_n$.

For each single-argument predicate $P(x)$, the state space S of the object x is divided into S^T and S^F, where S^T is the state space that makes $P(x)$ *true*, and S^F is the complement of S^T, that is $S^F = S - S^T$. Similarly, for each n-argument predicate $P(x_1, \cdots, x_n)$, the state space $S = S_1 \times \cdots \times S_n$ is divided into S^T and S^F which make the $P(x_1, \cdots, x_n)$ *true* and *false* respectively, that is

$$\langle x_1, \cdots, x_n \rangle \in S^T \Rightarrow P(x_1, \cdots, x_n) = true$$
$$\langle x_1, \cdots, x_n \rangle \in S^F \Rightarrow P(x_1, \cdots, x_n) = false$$

A pre- or post-condition of a use case model includes one or more predicates with objects, along with logical operators and quantifiers, and is determined *true* or *false* based on the value of each predicate that occurs in the condition. Therefore, the state space S of each condition can be divided into S^T and S^F in the similar way.

From the above discussion, a predicate based pre- or post-conditions can be transformed in the form of state based expression as follows.

1. If the predicate indicates a mathematical property like "=h, ">h, ">=h, and so on, transform it into a boolean expression like "$x_1 == x_2$", "$x_1 > x_2$", or "$x_1 >= x_2$" which makes the predicate *true*.
2. If the predicate indicates a non mathematical property, transform it into the form of $(x_1 == s_{11} \&\& x_2 == s_{21} \&\& \cdots) || (x_1 == s_{12} \&\& x_2 == s_{22} \&\& \cdots) || \cdots$ where $\langle s_{1i}, s_{2i}, \cdots \rangle \in S^T$.
3. For the quantifier "∀h, all the possible $(x_1 == s_{11} \&\& \cdots)$ are combined using "&&h, while for the quantifier "∃h, they are combined using "||h.

In a Promela code that expresses a use case, the initial state of each object must comply with the pre-condition of the use case. Therefore, at the beginning of the code, each variable must be assigned a value that makes the pre-condition *true*.

The correctness of a use case model that is expressed in Promela can be defned as "the final state of the code must satisfy the post-condition if the initial state satisfies the pre-conditionh. If there are multiple states $\{\langle s_1^{(i)}, \cdots, s_n^{(i)} \rangle\}$ that satisfy the pre-condition, and multiple states $\{\langle \sigma_1^{(j)}, \cdots, \sigma_n^{(j)} \rangle\}$ that satisfy the post-condition, the LTL formula to be evaluated would be

$$\Box((\mathbf{x} == \mathbf{s}^{(1)} || \cdots || \mathbf{x} == \mathbf{s}^{(n)}) \rightarrow \Diamond(\mathbf{x} == \sigma^{(1)} || \cdots || \mathbf{x} = \sigma^{(m)}))$$

where $\mathbf{x} = \langle x_1, \cdots, x_n \rangle$, $\mathbf{s}^{(i)} = \langle s_1^{(i)}, \cdots, \mathbf{s}_n^{(i)} \rangle$, and $\sigma^{(i)} = \langle \sigma_1^{(i)}, \cdots, \sigma_n^{(i)} \rangle$.

The above LTL formula must be evaluated at appropriate points in a Promela code. In order to control these points explicitly, we can introduce a *position marker* variable. This variable is assigned unique values at the above points. Assuming this variable is labeled "*pos*h, the statement "*pos* == a unique value" is appended to each term in the LTL formula.

The use case model shown in Table 2 is verified in the following way using the above process.

1. The objects that have states are "cashierh, "customerh, "producth, "cash in the registerh, and "sales amounth. The last two objects do not occur explicitly in the description, but are derived from it.
2. The following functions or methods can be regarded as *state sensitive*.
   - *receiveBasket*: affects the objects "cashierh and "customerh
   - *scan*: affects the object "producth
   - *getPrice*: affects the object "sales amounth
   - *showTotalAmount*: affects the object "cashierh
   - *receivePayment*: affects the object "cash in the registerh
   - *giveReceiptAndChange*: affects the objects "cashierh and "customerh

3. The states that are associated with the objects can be derived from the above methods as
   - { *isProcessed, isNotProcessed* } for a product
   - { *isReady, isProcesingItems, isProcessingPayment* } for a cashier
   - { *isWaiting, isInService, hasFinished* } for a customer
   - integer values for "cash in the registerh and "sales amounth
4. The initial state for the use case "Process an Itemh is expressed as $(cust == isWaiting) \&\& (item[i] == isNotProcessed)$ where *cust* and *item*[*i*] are the variables for a customer and an *i*th product in a shopping basket.

From these state conditions, we can derive the following Promela code. The code implements interconnected use case models. The proctype "*next1*()h connects the use case "Process an Itemh and "Paymenth based on the pre- and post-conditions relationships.

```
/*
        Use Case for Checkout System
*/
typedef product {
        int price;
        mtype state
}
typedef customer {
        mtype paymentOption;
        mtype state
}
mtype  = {isNotProcessed, isProcessed, isReady,
        isProcessingItems, isProcessingPayment, isWaiting,
        isInService, hasFinished, byCash, byCard};
mtype cashier;
customer cust;
product item[10];
product itemInTheCart[20];
int cashInRegister;
int salesAmount;
int pos;
int numberOfItems;
int selected;
active proctype prepare() {
        cust.paymentOption = byCash;
        numberOfItems = 5;
        salesAmount = 0;
        cashInRegister = 1000;
        int i = 0;
        do
        ::(i < numberOfItems) ->
                itemInTheCart[i].state = isNotProcessed;
                itemInTheCart[i].price = 10;
                i ++
```

```
        ::else  ->  break
        od;
        run customerArive();
}
proctype receiveBasket() {
        cust.state = isInService;
        cashier = isProcessingItems;
        run processItem(numberOfItems)
}
proctype processItem(int n)
{
        int i = 0;
        pos = 1;
        do
        ::(i < n) ->
        if
        ::(itemInTheCart[i].state==isNotProcessed) ->
                selected = i; break
        ::else -> skip
        fi;
        i++
        ::else -> break
        od;
        itemInTheCart[selected].state = isProcessed;
        salesAmount = salesAmount +
                itemInTheCart[selected].price;
        run next1()
}
proctype next1() {
        int unprocessedCount = 0;
        int j = 0;
        do
        ::(j<numberOfItems) ->
        if
        ::(itemInTheCart[j].state==isNotProcessed) ->
                unprocessedCount++
        ::else -> skip
        fi;
        j++
        ::else -> break
        od;
        if
        ::(unprocessedCount>0) ->
                run processItem(numberOfItems)
        ::else -> run payment(salesAmount)
        fi
}
proctype payment(int ammount) {
        cashier = isProcessingPayment;
        if
```

```
        ::(cust.paymentOption==byCash) ->
                cashInRegister = cashInRegister +
                        salesAmount;
        ::else -> skip
        fi;
        cashier = isReady
}
```

To this Promela code , the following LTL formulae can be defined in order to evaluate the validity of the use case model.

1. $\Box((P_1\&\&P_2) \rightarrow \Diamond(Q_1\&\&Q_2))$ where
   $P_1: \ cust == isWaiting$
   $P_2: \ item[0] == isNotProcessed$
   $Q_1: \ cust == isInService$
   $Q_2: \ item[0] == isProcessed$
2. $\Box((P_3\&\&P_4) \rightarrow \Diamond(Q_3\&\&Q_4))$ where
   $P_3: \ cust == isInService$
   $P_4: \ item[i] == isNotProcessed$
   $Q_3: \ cust == isInService$
   $Q_4: \ item[i] == isProcessed$
3. $\Box(P_5 \rightarrow \Diamond Q_5)$ where
   $P_5: \ item[numberOfItem] == isProcessed$
   $Q_5: \ cashier == isProcessingPayment$
4. $\Box(P_6 \rightarrow \Diamond Q_6)$ where
   $P_6: \ cust == isInService$
   $Q_6: \ cust == hasFinished$

and so on. These constraints to the model can be verified by the Spin model checker, and we can determine the model is valid.

5 Conclusions

A model formalization and verification process is introduced in this paper, which is dedicated to UML use case models. UML use case models are one of the most important models in object oriented software development, especially in a use case driven development. Therefore the correctness of these models is a critical issue for successful software development. However because of insufficient formalism in UML use case models, it is difficult to assure their correctness.

This paper formalizes use case descriptions based on first order predicate logic. The pre- and post-conditions are expressed in the form of PCNF logic formulae, while the basic and alternate courses of action are represented as a series of functions or methods. Even though these formalized models can depict use cases rigorously, there are few practical tools to assure the correctness of these logic based models.

For this assurance, the Spin model checker is applicable if the models represent state transitions. The paper proposes a systematic process to identify the states and their transitions in UML use case models in order to apply the Spin. Using this process, the logic based UML use case models can be transformed into Promela codes with which the Spin deals. Since the Spin needs LTL formulae to verify models, a procedure that creates them from the pre- and post-conditions is also proposed.

References

1. Ambler, S.W.: The Object Primer, 3rd edn. Cambridge University Press, New York (2004)
2. Ben-Ari, M.: Principles of the Spin Model Checker. Springer, New York (2008)
3. Elaasar, M., Briand, L.C.: An Overview of UML Consistency Management. Technical Report SCE-04-18, Carleton University (2004)
4. Fowler, M.: UML Distilled: A Brief Guide to the Standard Object Modeling Language, 3rd edn. Addison-Wesley Professional, Boston (2003)
5. France, R., Bruel, J.M., Larrondo-Petrie, M., Shroff, M.: Exploring the Semantics of UML type structures with Z. In: Proc. of the 2nd IFIP Workshop on Formal Methods for Open Object-Based Distributed Systems, pp. 247–260 (1997)
6. Holzmann, G.J.: The SPIN Model Checker: Primer and Reference Manual. Addison-Wesley Professional, Boston (2003)
7. Jacobson, I.: Object-Oriented Software Engineering: A Use Case Driven Approach. Addison-Wesley Professional, Boston (1992)
8. Kuske, S.: A formal semantics of UML state machines based on structured graph transformation. In: Proc. of the 4th International Conference on the Unified Modeling Language, pp. 241–256 (2001)
9. Pilone, D., Pitman, N.: UML 2.0 in a Nutshell, 2nd edn. O'Reilly, Sebastopol (2005)
10. Rosenberg, D., Stephens, M.: Use Case Driven Object Modeling with UML: Theory and Practice. Apress, Berkeley (2007)
11. Shinkawa, Y.: Inter-Model Consistency in UML Based on CPN Formalism. In: Proc. of the 13th Asia Pacific Software Engineering Conference, pp. 411–419 (2006)
12. Wieringa, R.: Formalizing the UML in a Systems Engineering Approach. In: Proc. of the 2nd ECOOP Workshop on Precise Behavioral Semantics, pp. 254–266 (1998)

Author Index